World Soils Book Series

Series editor

Prof. Alfred E. Hartemink
Department of Soil Science, FD Hole Soils Laboratory
University of Wisconsin–Madison
Madison,
USA

Aims and scope

The World Soils Book Series brings together soil information and soil knowledge of a particular country in a concise and reader-friendly way. The books include sections on soil research history, geomorphology, major soil types, soil maps, soil properties, soil classification, soil fertility, land use and vegetation, soil management, and soils and humans.

International Union of Soil Sciences

More information about this series at http://www.springer.com/series/8915

José Espinosa · Julio Moreno
Gustavo Bernal
Editors

The Soils of Ecuador

Springer

Editors
José Espinosa
Facultad de Ciencias Agrícolas
Universidad Central del Ecuador
Quito
Ecuador

Gustavo Bernal
Asociación Nacional de Cultivadores de Palma Aceitera (ANCUPA)
Quito
Ecuador

Julio Moreno
Instituto Espacial Ecuatoriano (IEE)
Quito
Ecuador

ISSN 2211-1255 ISSN 2211-1263 (electronic)
World Soils Book Series
ISBN 978-3-319-25317-6 ISBN 978-3-319-25319-0 (eBook)
https://doi.org/10.1007/978-3-319-25319-0

Library of Congress Control Number: 2017952925

Printed on acid-free paper

This Springer imprint is published by Springer Nature
The registered company is Springer International Publishing AG
The registered company address is: Gewerbestrasse 11, 6330 Cham, Switzerland

Preface

Due to the influence of the Andes, which vastly modified climate and parental materials, a relatively small country like Ecuador has a wide variety of soil orders, hard to find in other countries of the world at this scale. In spite of its great diversity and importance, soil is perhaps the least known natural resource in Ecuador. Soil maps at different scales, and their respective proceedings, have been published over the years, but unfortunately, this information has been kept in the hands of few institutions and individuals and has not been available to general public. Currently, enough information is available to describe the fascinating and unique characteristics and distribution of Ecuadorian soils. With all of this in mind, the book Soils of Ecuador trays to organize, synthesize, and socialize the generated knowledge about Ecuadorian soils. The book is based on the soil maps developed during the 1970s and 1980s by the Regionalization National Program (PRONAREG) and ORSTOM (currently IRD). In 2002, these maps were digitalized and geo-referenced by The Center for Integrated Surveying of Natural Resources by Remote Sensing (CLIRSEN), currently Ecuadorian Spatial Institute (IEE), the Minister of Agriculture, Livestock and Fisheries (MAGAP), and the Inter-American Institute for Cooperation in Agriculture (IICA). All this information was complemented by the information developed by the project called "Generation of Geo-information for the Development of the Ecuadorian Territory" financed by the Ecuadorian Government and conducted by IEE, MAGAP, and The National Secretariat for Planning and Development (SENPLADES).

This book is organized to represent the following concepts : (1) landscapes, as natural integrated geographic systems conceived as complex zones resulting from structural arrangement of the relief related by proximity, and complexity, and characterized, to certain extent, by its morphogenetic and pedogenetic behavior; (2) regions, as an integration of landscapes defined by the following strict rules: (a) ample enough, (b) clearly defined original physical characteristic of any nature, and (c) physical and geographic conditions easily differentiated from the surroundings; and (3) Ecuadorian natural environments as the integration of the regions based on physical reality of the country and the perception of them by the Ecuadorian population.

The writing of the Soils of Ecuador book is the result of the joint effort made by experienced professional representing the Central University of Ecuador—College of Agriculture, the Ecuadorian Spatial Institute, and the Ecuadorian Associations of oil Palm Growers, in coordination with a group of young soil scientist who have developed important field surveying work all over the country. The result of this coordinated initiative is a detailed description of the soils from this small and widely diverse country in the center of the world.

Quito, Ecuador José Espinosa
Quito, Ecuador Julio Moreno
Quito, Ecuador Gustavo Bernal

Contents

1 Introduction

Julio Moreno, Gustavo Bernal, and José Espinosa

Abstract

The Andes divides Ecuador in three natural regions: Coastal Plane, Highlands and Amazonia, each one a complex and varied system of climate, landscapes, soils, and biodiversity product of the equatorial conditions of the country and the altitude. For these reasons, sunshine is not uniform in all areas of the country promoting the presence of anarchic winds and distinct pluvial regimes that controlled soil development. The Galapagos Islands are also included in the national territory. These islands emerged from the Pacific Ocean due to the eruption of submarine volcanos developing landscapes, also influenced by climate and altitude. The Coastal Plane has an extended plain of high fertility soils but step slopes around the central coastal cordillera limit agricultural production and the humid section of the north has poor ferralitic soils. The highlands, demarcated by the two cordilleras, comprise a zone from 1300 to 2800 m a.s.l. This region has the steep slopes as the main limiting factor for agriculture. The northern section of the Highlands has richer soils due to volcanic ash contribution. The center and southern sections of the Highlands do not have volcanic ash cover from volcano activity, but climate and parental materials have promoted the development of a variety of soils, but there are many areas dominated by poor fersalitic soils. In the Amazonia region, the areas close to the cordillera have alluvial or volcanic ash-derived soils which can be used for agriculture if slope is not limiting. The rest of the Amazonia has alluvial hydromorphic or poor ferralitic soils forming a particularly fragile area with severe limitations for agriculture. Forest and all environmental services of this zone have to be conserved. The Galapagos Islands has mineral soils due to the arid climate, with the exception of the humid volcano flanks where the alteration of pyroclastic materials has developed Andisols, while the lava accumulation has given place to fersalitic poorly evolved soils. Only a small section of the Galapagos Islands are used for agricultural activities due to restriction to protect the national park.

Keywords

Ecuador • Coastal Plane • Highlands • Amazonia • Galapagos Islands

1.1 Historical Review of Soil Surveying

The earliest studies of the Ecuadorian soils were the Coastal Plane and Highlands descriptions conducted by Hugh H. Bennett in 1922. These non-published studies were used by Misael Acosta Solis to publish the first Coastal Plane and Highlands map of Ecuador (Acosta Solís 1965; Pacheco 2009).

Between 1940 and 1952, Misael Acosta Solis carried out geo-botanic studies in several provinces of the country. The

J. Moreno (✉)
Instituto Espacial Ecuatoriano, Quito, Ecuador
e-mail: julio.moreno@institutoespacial.gob.ec

G. Bernal
Asociación Nacional de Cultivadores de Palma Aceitera (ANCUPA), Quito, Ecuador
e-mail: gbernal@ancupa.com

J. Espinosa
Facultad de Ciencias Agrícolas, Universidad Central del Ecuador, Quito, Ecuador
e-mail: aespinosa@uce.edu.ec

J. Espinosa et al. (eds.), *The Soils of Ecuador*, World Soils Book Series,
https://doi.org/10.1007/978-3-319-25319-0_1

objective of these land and native vegetation studies was to define a group of indicator plants of each edaphic group, but not all of these studies were published. Acosta Solis also participated in several excursions made by American scientists to the three natural regions of Ecuador. The most relevant of these studies was the work conducted by Robert L. Pendleton and Charles S. Simmons, which were complemented by Miller and used to classify most of Ecuadorian soils within the internationally recognized large groups and to create new names, especially those for the páramos. The compilation of all these studies was published by Miller in 1945 in his map of the main zonal groups of soils of Ecuador (Acosta Solís 1965; Ortiz et al. 2010; Pacheco 2009).

In 1955, a technical assistance agreement was signed between the Food and Agricultural Organization of the United Nations (FAO) and the government of Ecuador to begin soil surveying and classification studies. Several studies at the exploratory level in almost all the national territories were conducted with the support of Erwin Frei, FAO consultant, who worked in the country in 1955 and 1956, to develop the first inventory and determine the geographic distribution of soils of the country. These studies detected the existence of 21 different great soil groups in Ecuador. Frei was replaced by Alfredo Kupper, who continued with the national plan of soil surveying and classification (Acosta Solís 1965). The collected information was mapped in the soils map of Ecuador, scale 1:1´000,000, the only document of this type available until that date and the one which served as the basis to draw the general soils outline of the country reduced to a scale of 1:2´000,000 (CLIRSEN 1979).

During the 1960s, several specific soil studies were conducted. One of them was the Preliminary Study on Ecology, Soil and Agricultural Potential of the Eastern Region of Ecuador made by Rafael Pacheco, who presented a 1:500,000 map which includes six major soil associations with names, location and separation. Other soil survey was conducted for the Center for Conversion of Azuay, Cañar and Morona Santiago (CREA) by Alfredo Kupper, Rafael Pacheco, and Víctor Andrade. A report and the map at 1:500,000 scale was presented, recognizing the presence of six great soil groups, two associations, and one transition in the territory of those provinces (Acosta Solís 1965).

In the 1970s, Ecuador experienced a radical change with the oil discovery and exploitation in ample territories of the Amazonia. This situation generated abundance of monetary resources which, under the slogan "of seeding the oil," allowed the government to give a great impulse to the economic activities of the country, particularly those of the agricultural and livestock sectors. This way, and with good vision, support was given to develop the Agricultural Map of Ecuador. The ministry of Agriculture and Livestock (MAG) was on charge of the implementation of the studies and of the map elaboration.

With the goal of providing enough technical consistency to the Ecuadorian agricultural planning, the Ministry of Agriculture and Livestock created, in 1973, the National Agricultural Regionalization Program (PRONAREG), which was to be on charge of studying and evaluating the renewable natural resources and to conduct agro socioeconomic research. Agronomists with experience in soil surveying from several other government institutions were the core of the human talent of PRONAREG. The first PRONAREG initiative was to organize a national course on norms, methodologies, and procedures to conduct soil surveying (González et al. 1986). Part of this pioneer group of soil scientists were Fausto Maldonado, Jacinto Yépez, Néstor Espinosa, Gustavo Peña, Carlos Luzuriaga, Héctor Loor, Patricio Toledo, Augusto González, and Richard Guillén.

In 1974, PRONAREG signed an agreement for technical cooperation with the *Office de la Recherche Scientifique et Technique Outre Mer* (ORSTOM), but the actual field work was initiated in February 1975 due to time needed to collect basic data to support the complete analysis of the Ecuadorian rural environment (Portais 1997). At the begging of this relevant ORSTOM contribution, among the most important, the following names of French researchers can be mentioned: Francoise Colmet Daage, Michael Portais, Pierre Pourrut, Alain Winckel, Claude Zebroswki, George De Noni, Michael Sourdat (González 2015).

During the period between 1976 and 1986, the PRONAREG-ORSTOM project conducted 16 soil surveys in Ecuador, generating morpho-pedologic maps of the Amazonia (scale 1:500,000), Coastal Plane (scale 1:200,000) and Galapagos Islands (scale 1:50,000), and Highlands soil maps at 1:50,000 and 1:200,000 scale. The taxonomic classification was generalized to the Great Group level and in some cases to the Subgroup level, using the North American Soil Classification System (Soil Survey Staff 1975). As a complementary tool for describing soil profiles, the FAO Guide for Soil Profile Description was also utilized (Mejía 1997; Ortiz et al. 2010). These soil studies were not only based on soil data, but also in geomorphology studies supported by photo interpretation and laboratory analysis of soil physical, chemical, and mineralogical characteristics conducted in laboratories of Ecuador, the Antilles and France. Although the basic work was done at various scales, finally they were presented at the scales indicated before (Portais 1997).

The profile data sheets and laboratory reports produced by the PRONAREG-ORSTOM project included 19 physical-chemical soil attributes, four geomorphological attributes, and two climatic attributes. The details of the data sheets include location (region), altitude, landscape, slope, precipitation, atmospheric temperature, macro-nutrients, exchangeable cations, pH (water and KCl), NaF and carbonates reaction, carbon/nitrogen ratio (C/N), humic and fulvic acid content, horizon depth, color, mottling, texture, structure, soil moisture and temperature regimes, bulk density and available water (Ortiz et al. 2010).

During this period, other soil surveys were also conducted at semi-detail and detailed level. Within the first, 37 projects were generated, among them the Sierra soil and slope map at 1:50,000 scale which was generated by the PRONAREG-ORSTOM project. At the detailed level, 18 projects were conducted by institutions such as the Ecuadorian Institute of Water Resources (ex-INHERI), the Manabí Rehabilitation Center (ex-CRM), Commission for the Development of the Guayas River Basin (ex CEDEGE) and to a lesser extent by PRONAREG (CLIRSEN 1979; Mejía 1997).

In 1974, soil studies from South America, including Ecuador, were compiled. This project grouped soil in sections according to the management and discusses methods to correlate the different classification systems for land use, fertility limitations, acidity and fertility management, and a mechanism to develop a research network of tropical soils (Bornemiza and Alvarado 1974).

Several regional soil surveys were conducted in this decade. The most important was the General Exploratory Soil Survey for the Loja and El Oro provinces conducted by MAG in 1975 covering a surface of 13,767 km^2. The results were published at 1:250,000 scale including soil mapping units in associations and consociations. In 1977, the Program for the Development of the South (PREDESUR) developed a soil survey and its agrologic interpretation supported by the Puyango-Tumbez binational cooperation project (Ecuador-Peru). The survey covered a surface of 69,016 ha in areas considered as priority for Ecuador. Detailed soil surveys for intensive agricultural use, specific development projects, and establishment of experimental farms, cadastral evaluation at farm level, land productivity, irrigation channel construction and global water needs were conducted by the Ecuadorian Institute of Water Resources (INERHI), CRM, CEDEGE, and MAG (González 2015).

Ecuador and Chile were the venue for the Sixth International Soil Classification Workshop in 1984. The andic properties that characterize soils derived from volcanic ash were defined in this event, properties that later were documented by the International Committee for the Andisols classification (ICOMAND) and that served as the basis for defining the new order of Andisols in the Keys for Soil Taxonomy in 1990. Before that date, the soils derived from volcanic ash were included in the Inceptisols Order as the suborder Andepts (Dahlgren et al. 2004; Shoji et al. 1993; Soil Survey Staff 1990).

In 1986, the Ecuadorian Soil Science Society (SECS) published, at a 1:1′000,000 scale, the General Soil Map of Ecuador, based on the soil cartography previously generated by the PRONAREG-ORSTOM project at the 1:200,000 scale for the Highlands and Coastal Plane, 1:500,000 for the Amazonia region, and 1:50,000 for the Galapagos Islands. Thirty six Great Groups were identified in this map, grouped into 18 suborders and 8 orders, according to the classification system of the United States Department of Agriculture (USDA) (González et al. 1986).

Mejia, in 1997, elaborated the General Soil Capacity—Fertility Map of Ecuador (scale 1:2′000,000), based on the same PRONAREG-ORSTOM soil surveys used to generate the General Soil Map (1:1′000,000). This map allows the interpretation of soil characteristics as a guidance to perform preliminary diagnosis of similar soil fertility, agriculture, livestock, and forest areas of a particular location in Ecuador (Mejía 1997).

On the other hand, in 1997, two books were written based on information generated by the PRONAREG-ORSTOM project and were published as a joint effort of the American Institute of Geography and History (IPGH), the Military Geographic Institute (IGM), ORSTOM and the Ecuadorian Geographical Research Center (CEDIG). One of them was titled General Conditions of the Natural Environment (Winckell et al. 1997a). This book discusses the main aspects of the relief and geodynamic evolution of the country, including mineralogy and internal forces that acted for its formation, analysis of surface processes, major climatic regions, natural vegetation, hydrographic and hydrological features, pedogenic factors and aspects on soil erosion. The second book, titled Regions and Landscapes of Ecuador (Winckell et al. 1997b), defines the landscape as the result of the variable combination of physical, biological, and anthropogenic components that interact with each other, forming a comprehensive set of constant change, where soil is part of the landscape. In this book, the landscape discussion is supported by the description of the general conditions of the most important natural components, considering the country's geography at the level of morpho-pedological units and, collectively, as forms of the regional topography.

It is important to mention that in 1996, the III International Symposium on Hardened Volcanic Soils took place in Ecuador with the objective to address the problem of cangahuas and tepetates from Ecuador, Colombia, and Mexico (Zebrowski et al. 1997). These soils are located at the volcano foothills of very populated areas that have been affected by constant erosion due to inappropriate farming practices. The proceedings of the event presents the results of studies on the characteristics of hardened materials, factors affecting productivity, erosion and conservation practices, and socioeconomic conditions for their rehabilitation for agricultural and agroforestry use.

The Center for Integrated Surveying of Natural Resources by Remote Sensing (CLIRSEN), institution on charge of the natural resources inventory in Ecuador, changed its name to Ecuadorian Space Institute (IEE) in 2012. During the period from 2009 to 2014 CLIRSEN-IEE, in cooperation with the Ministry of Agriculture, Aquaculture and Fisheries (MAGAP) and the National Secretariat of Planning and Development (SENPLADES) evaluated 4031 soil profiles within the project called Generation of Geo-information for Planning at National Level (IEE 2014). The objective of this study was to generate geo-information of the country at a 1:25,000 scale as a tool for land management, risk management, and sustainable improvement of agricultural production. The need for updated information about the most important geological, soil, hydrological, climatic and socioeconomic conditions of the country compelled the government of Ecuador to declare this project as a priority for national development (MIDENA et al. 2013).

In 2014, a group of the IEE professionals participated in the development of the Soil Atlas of Latin America and the Caribbean, which presented updated information on the soils of Ecuador (Gardi et al. 2014). The taxonomic information of the General Map of Ecuador at 1:1′000,000 scale, published in 1986 by the Ecuadorian Soils Science Society, was updated as part of this work using the World Reference Base of the Soil Resource (WRB) as the classification system.

This book, titled Soils of Ecuador, is the result of the combined effort of several Ecuadorian specialists, who, with the support and encouragement of Juan Gallardo Lancho and Christian Prat, scientists with the PROMETEO project of Ecuador, gather at the end of 2013 to present a proposal to the Springer publisher for the edition of the first book on this subject in the country. The work is based on soil maps developed during the 1970 and 1980 decades by the PRONAREG-ORSTOM (now MAGAP and IRD) project, maps that were digitized and geo-referenced by CLIRSEN, MAGAP and IICA in 2002, as well as the field work (profile description) conducted to generate geo-pedologic maps, at a 1:25,000 scale, by IEE, SENPLADES, and MAGAP, within the government project called Generation of Geo-information for Planning at National Level.

1.2 Geographic Aspects

The Republic of Ecuador (mainland and islands) is located in the Western Hemisphere, at northwestern side of South America (Fig. 1.1).

Ecuador continental territory is located between the latitudes 01°28′N and 05°02′S and between longitudes 75°11′W and 81°04′W. The insular territory or Galapagos Islands is located 1000 km west of mainland between latitudes 1°39′N and 1°26′S and between longitudes 89°15′W and 92°01′W. The country has a total area of 256,370 km^2, bordered by Colombia to the north, Peru to the south and east, and the Pacific Ocean to the west. The Equator, or parallel 0°, crosses the country bisecting the continental and insular territories, but most of the territory is located in the Southern Hemisphere (IGM 2013).

1.3 Regions and Natural Landscapes

The Andes mountain range divides the continental territory in three different regions: Coastal Plane, Highlands and Amazonia, each of them a complex and diverse system of climates, soils, biodiversity, and landscapes present due to the combination of the equatorial character of the area and the altitude (sea level to 6310 m at the summit of Chimborazo volcano) (Fig. 1.2). Due to these conditions, sunshine is not uniform throughout the country, producing anarchic winds and different rainfall patterns. The Insular Region consists of several islands rising from the Pacific Ocean by successive underwater volcano eruptions that formed the actual landscapes which also developed conditions influenced by altitude (Fig. 1.3) (IGM 2013).

The Coastal Plane region extends from the shoreline (sea level) to the western slope of the Andes reaching an altitude of 1200 m. The Highlands region is formed by the western and eastern Andes mountain ranges, arranged in meridian direction, with very steep outer slopes which are characterized by a general decline of altitude and massiveness from north to south (IGM 2013).

The Amazonia region extends from the foot of the Andes eastern slope, but between 500 and 1500 m over the sea level a strip 50 km wide that resembles a third cordillera is present. Then, below 300 m, the Amazonia plain materilizes reaching the eastern boundary with Peru. These plains consist of a monotonous landscape formed by small hills about 50 m high. Finally, the Insular region, consisting of 13 major islands, is located at a distance between 900 and 1200 km from the mainland. These islands were formed by volcanoes emerging from the sea, have gentle slopes and calderas culminating at about 1600 m (IGM 2013).

Within each of these regions it is possible to find different landscapes (Winckell et al. 1997b). This way, the Highlands are made of the following formations (Fig. 1.4): (a) Andean slopes and ridges, (b) Inter-Andean basins of the northern Highlands with pyroclastic cover, (c) Basins and valleys of the central Highlands without pyroclastic coverage, and (d) Reliefs of the southern Highlands.

On the Coastal Plane, the following landscapes are observed (Fig. 1.5): (a) Northwestern coastal regions, (b) Central coastal regions of Manabí, (c) Salinas peninsula and Isla Puná, and (d) Central plain and its foothills.

In the Amazonia the landscapes present are as follows (Fig. 1.6): (1) Sub-Andean reliefs corresponding to the side close to the Andes, and (2) Peri-Andean Amazonia.

Fig. 1.1 Geographic location of Ecuador (IGM 2013)

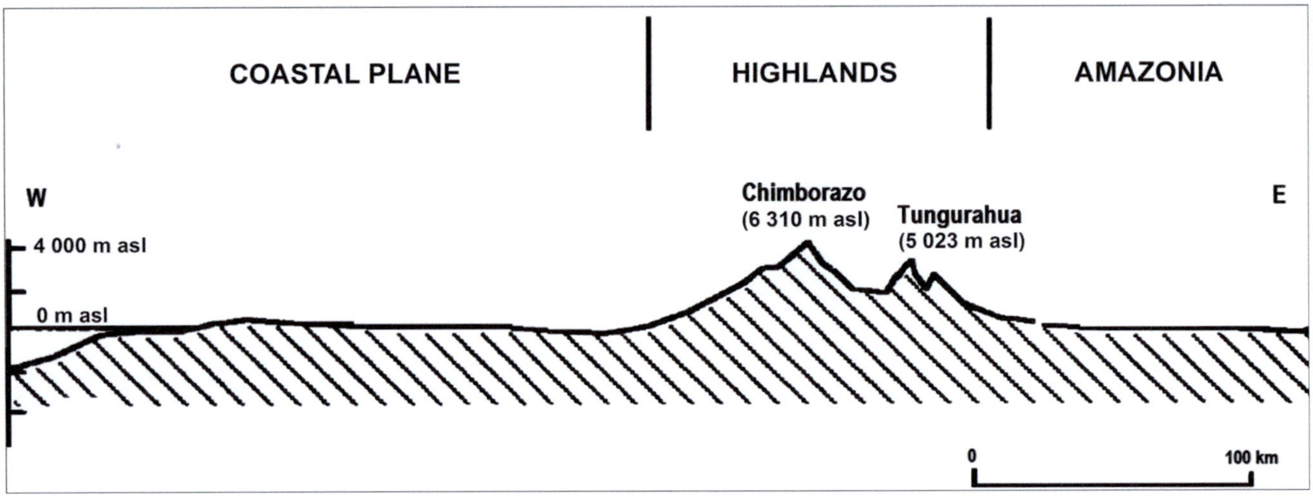

Fig. 1.2 Schematic configuration of the topography of continental Ecuador (Marocco and Winter 1997)

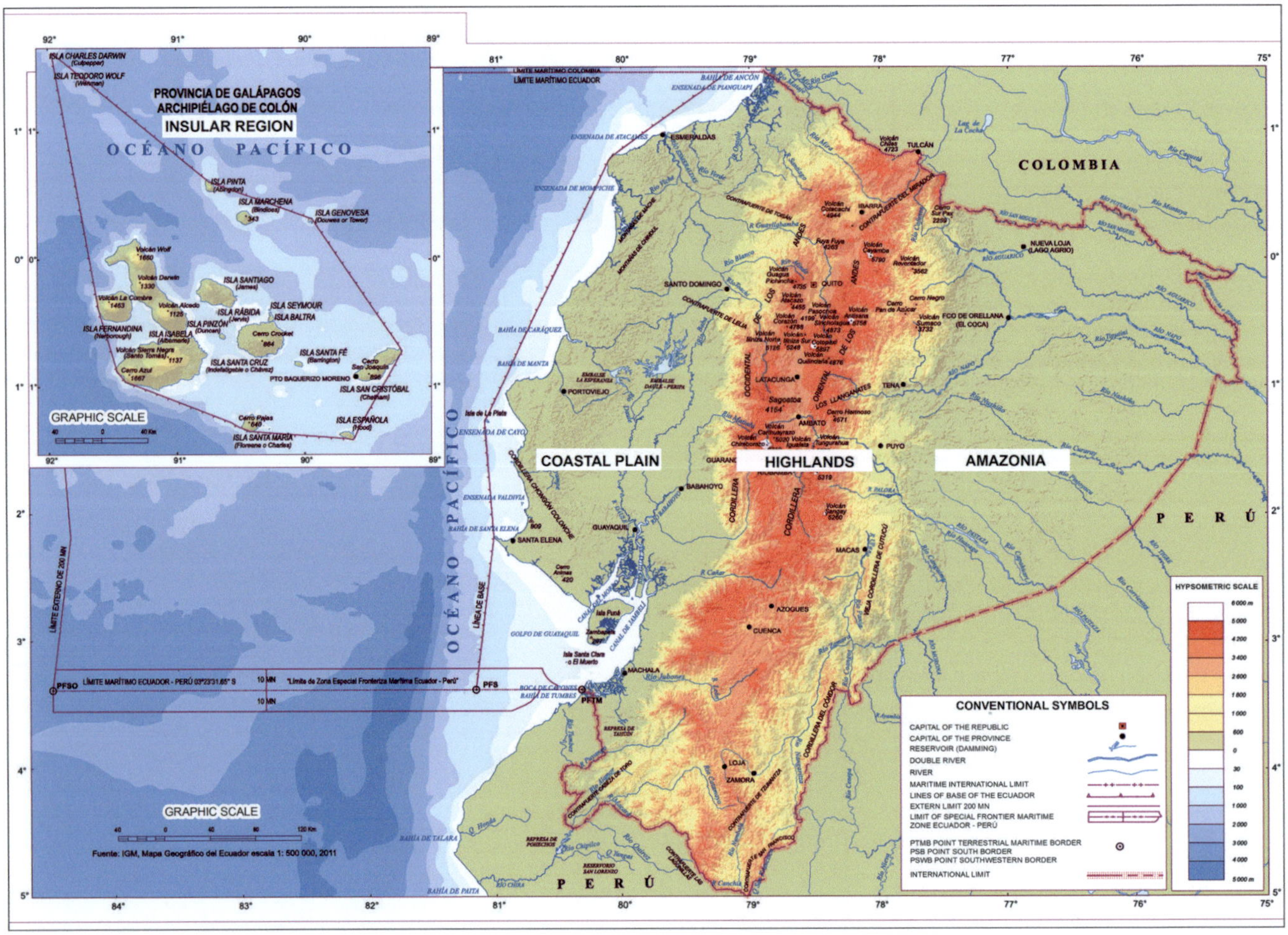

Fig. 1.3 Map of Ecuadorian topography (IGM 2013)

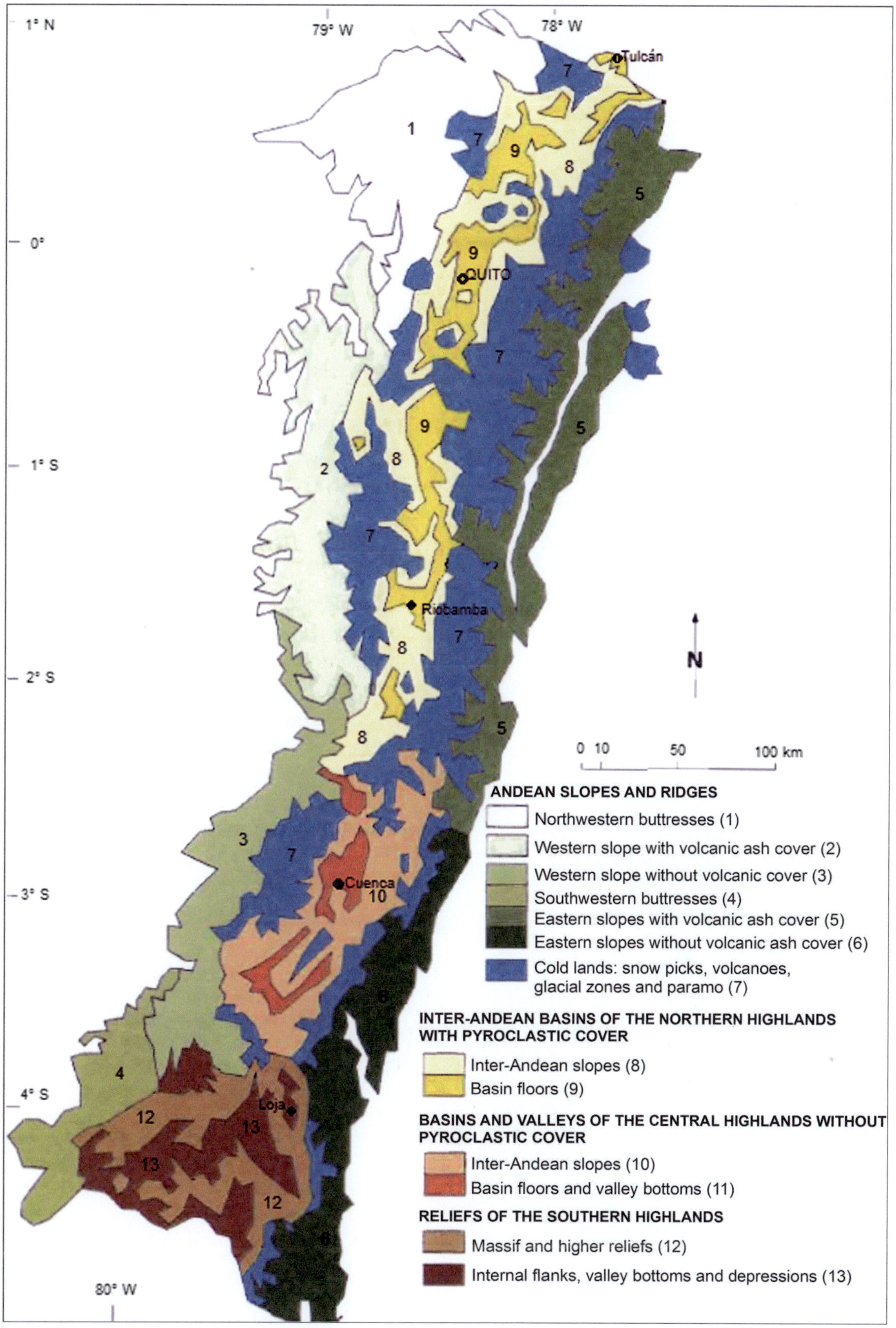

Fig. 1.4 Highlands landscapes (Winckell et al. 1997b)

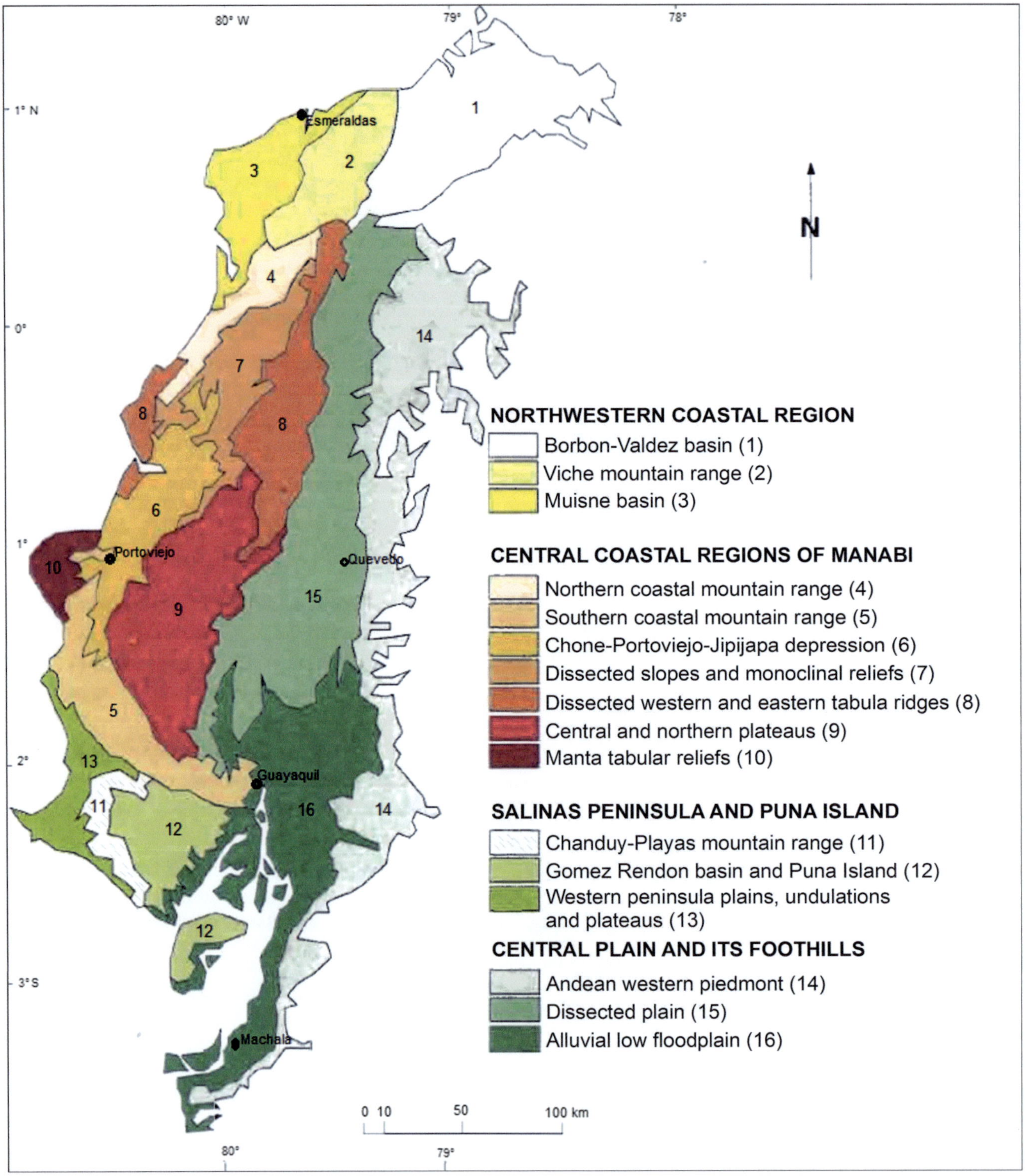

Fig. 1.5 Coastal plane landscapes (Winckell et al. 1997b)

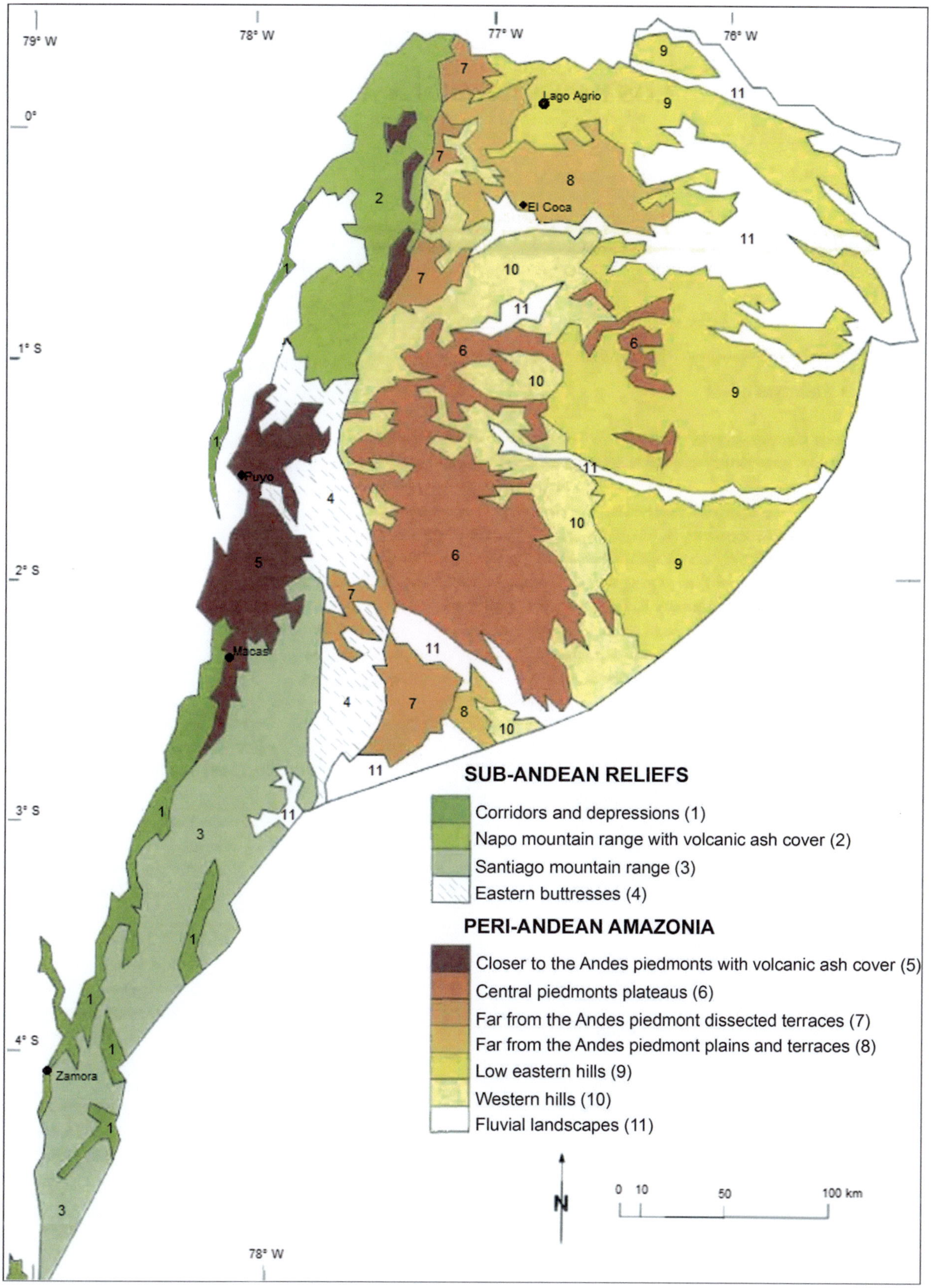

Fig. 1.6 Amazonia landscapes (Winckell et al. 1997b)

Finally, the landscapes of the Galapagos Islands are (Fig. 1.7): (1) Low islands, (2) Shield volcanoes without caldera, (3) Shield volcanoes with caldera, and (4) Common forms of the islands as a whole.

1.4 Reliefs

The presence of the Andes, the great meridian mountain range located in the middle part of the continental Ecuador, separates the territory into three regions of indisputable boundaries and definitions. The Andes mountainous relief is located in the center, the Coastal hills and plains to the west, and the Amazonia ridges, hills, and plains to the east (Fig. 1.8). The Galapagos low islands and high active volcanoes with caldera are the fourth region of the country.

1.4.1 Highlands Reliefs

The Andean Highlands or Sierra is the most notorious mountain range of Ecuador, with a width that varies from

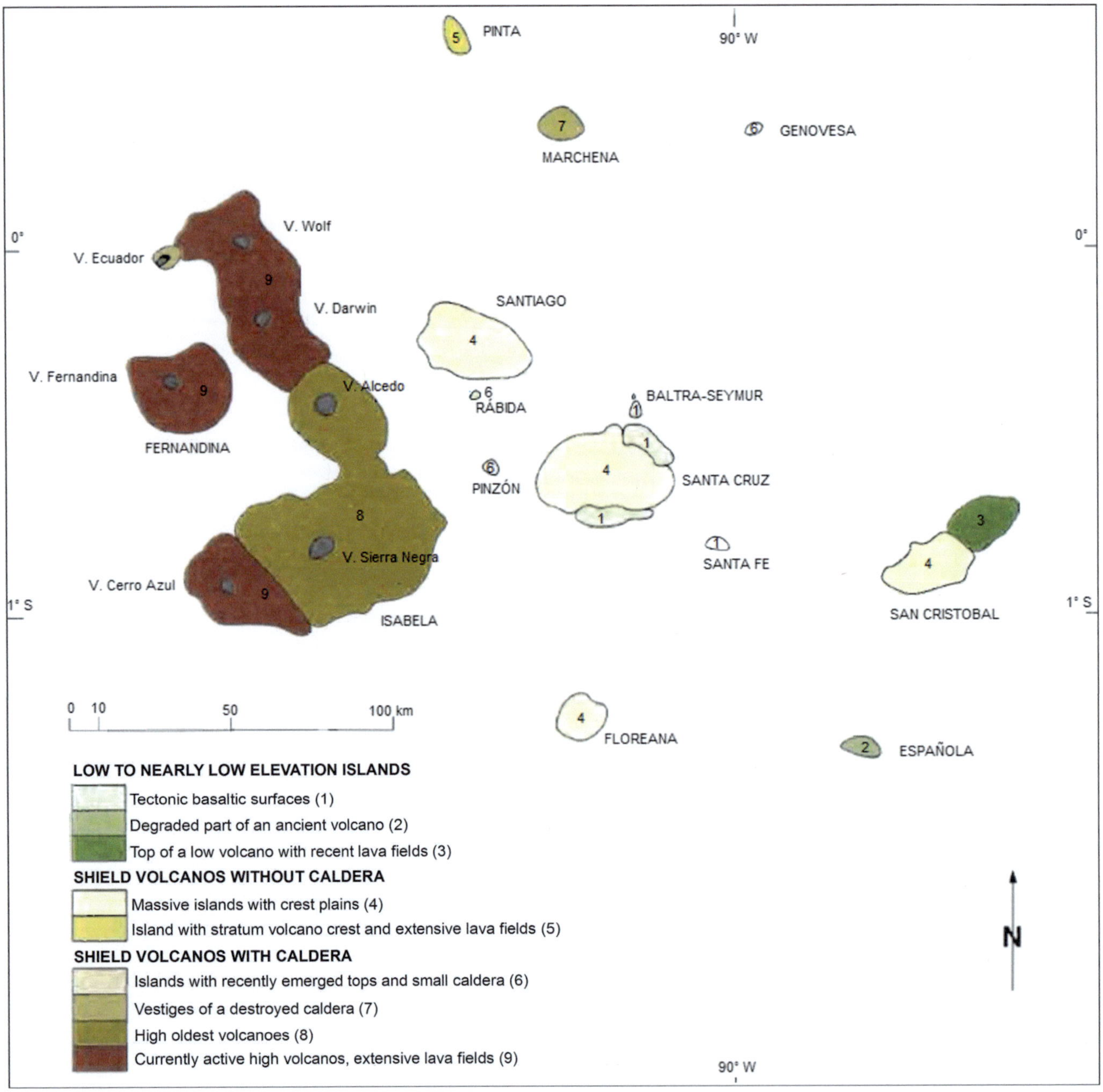

Fig. 1.7 Galapagos Islands landscapes (Winckell et al. 1997b)

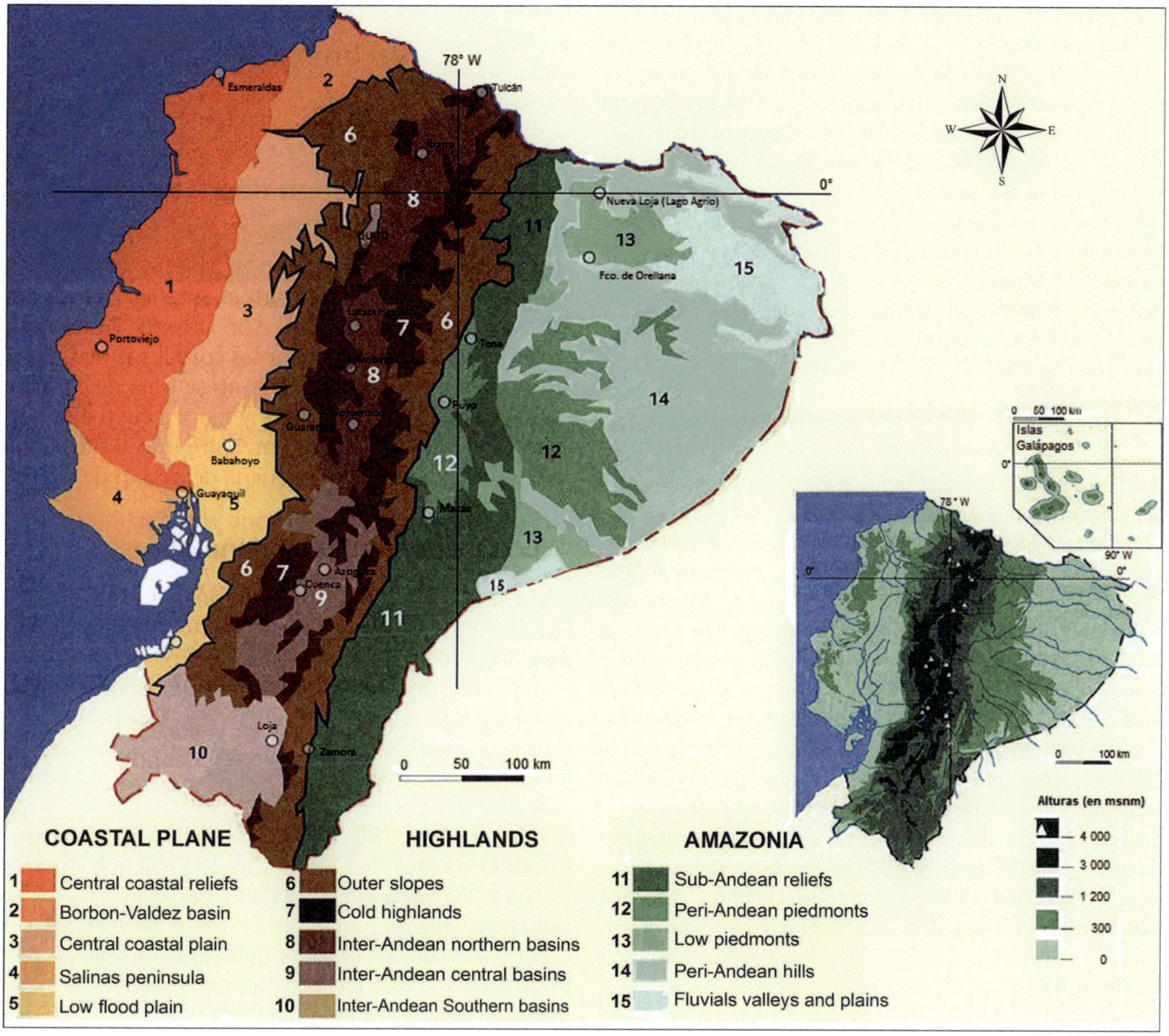

Fig. 1.8 Regions of continental Ecuador (León 2010; Winckell et al. 1997b)

100 to 150 km from east to west. The range central part is the narrowest, but it widens to the south due to the presence of Amazonia Sub-Andean mountain ranges reaching a width of 180–200 km. In the north, in the area located between the border with Colombia and the Equator, the range has a SSW–NNE and SW–NE direction which extends into Colombia. In the center, between the Equator and parallel 2° 30′S, the direction is meridian. To the south, between the parallel 2° 30′S and the border with Peru, the Andes are oriented in a SSW–NNE direction, parallel to the coastal line (Winckell et al. 1997b).

Northern Andes

From the Colombian border to latitude 2°15′S (Palmira-Alausí), the great mountain range forms what Humboldt called the "Avenue of the Volcanoes" due to the presence of several volcanoes, some active like Tungurahua with 5023 m, and other not active like Chimborazo with 6310 m, which form two secondary ranges (cordilleras) both sides of the Andes (Winckell et al. 1997b). These cordilleras have very steep external flanks that are visible from the coast or from the Amazonia, with elevations up to 4000–4500 m, meridian orientation in the center, and curved to the east on the northern edge.

The relief of this area is characterized by the quaternary glacier footprint found on the summits and for the frequent eolian deposits of volcanic materials. Between the cordilleras is the Inter-Andean Basin which is a depression just 20 km wide in the central part. This depression is divided transversally by knots formed by coalescing volcanoes developing internal basins of significant extensions, such as the basin of Quito or El Chota Valley, that generally have their own hydrographic system. These basins stand over a floor of volcano-sedimentary fillers at altitudes that vary from 1600 to 3000 m (Winckell et al. 1997b).

Central Andes

The Central Andes extend to the south from Palmira-Alausí as far as the latitude of Zaruma-Saraguro. This section of the Ecuadorian Andes is shaped as a wide a monotonous plateau with no recent large volcanoes as in the Northern Andes. The relief of this area was formed by volcanic outpourings which in turn influenced the geomorphology. The width of the platform reaches 40 km west of Cuenca and the altitude decreases towards the south where it only reaches 3800 m west of Zaruma. The two main parallel valleys, oriented NE to SW, vary in altitude from 1600 m at Santa Isabel to 2500 m at Saraguro, and they join together through an orthogonal mountainous section at the east. This way, Cuenca, Girón, and Santa Isabel are located to the west and Gualaceo, Nabón, and Saraguro to the east. These valleys open to the Coastal Plane through the Huigra, Cañar, and Jubones rivers (Winckell et al. 1997b).

Southern Andes

From the latitude of Zaruma-Saraguro to the Peruvian border, the relief is hazy and characterized by the presence of low sites that go SSW and higher sites which barely reach 2500 m and show no evidence of glaciers (Winckell et al. 1997b). The relief in the south is formed by elongated ridges where enlarged transverse valleys, as those formed by the Catamayo and Puyango rivers, that drain small isolated depressions which have different shape. The fragmented aspect of the Southern Andes is notorious for the progressive reduction of altitude of the Western Cordillera which gradually changes to form the lower reliefs of the Western Coastal Plane.

1.4.2 Coastal Plane Reliefs

The Coastal Plane starts at the foot of the western Andean range with a north to south orientation running parallel to the Andean Highlands. The width of this strip is variable, reaching about 100 km at the north, 180 km at the center, while at the south it decreases to 20–40 km around Guayaquil. In general, the relief of the Coastal Plane is moderate; the highest point reaches 830 m and is located in the southcentral section of the region. The Coastal Plane has two distinct geographic sections: (a) Coastal reliefs from the west and north and (b) Coastal flat lands from the east and south (Winckell et al. 1997b).

Western Coast Reliefs

This area is located to the west in a line connecting Quinindé (north) with Guayaquil (south). These reliefs are divided into three well-defined parts: North Coast, Central Coast, and Salinas Peninsula (Winckell et al. 1997b).

North Coast: Between the Guayllabamba river and the border with Colombia, there is a large round basin surrounded by high neighboring relieves. The southeast side of this basin is the western flank of the Andes, and the other side is made by the high hills of Viche, which are the north end of the Central Coastal relives, with altitudes ranging from 400 to 600 m.

This great depression, open to the Pacific Ocean to the northwest, has a very light relief, dominated by tabular structures and dissected hills with altitudes between 200 and 400 m, which in the Borbon-Valdez basin become small hills and rolling plains.

Central Coast: This formation has high reliefs characterized by a center line 20 km wide and 800 m high called Mache to the north and Colonche to the south. This mountain range, with an overall NNE–SSW orientation, takes a curve to the southeast (south of Jipijapa) and continues down to Guayaquil. These reliefs disappear in the middle of the range, between Chone and Portoviejo section, where only a few massifs of about 600 m are present.

The Central Coast is bordered on both sides by hilly reliefs. A narrow strip of fragmented low altitude hills (400 m) is located to the west, with exception of the relief located at the southwest of Manta that advances to San Lorenzo, the relief between Portoviejo and Chone, with river valleys, and the relief located north of Muisne. To the east, the relief is simple and consists of two physical ensembles: (a) Relief with sedimentary features located north, in a line from Portoviejo to El Empalme which slowly decreases in altitude towards the east and leans to the eastern plain, and (b) Southern relief of attractive horizontal tabular formations 400–450 m high, oriented towards Guayaquil.

Salinas Peninsula: This Peninsula, located south of the Colonche mountain range, is more complex and consists of three different zones arranged from east to west. At the center, in a SE–NW direction, there exists a small narrow range spitted in isolated massifs which ends at 420 m in

Chanduy. This small cordillera separates notoriously the littoral plains and low plateaus (less than 150 m) from the hilly relieves of Gómez Rendón basin in the east. The relief gradually lowers to the east reaching the Gulf of Guayaquil mangroves. The Puná Island has the same relief, but with lower altitude, the highest point reaches 260 m.

Western Coast Plains

These plains extend from Quinindé at the north to Huaquillas at the south, reaching a width of 90 km, from east to west, in Guayaquil, and 15–20 km north of Machala. There are two types of plains: (a) Central High Plains, (b) Low Alluvial Plain.

Central High Plains: This undulating relief is located between Quinindé at the north and Daule-Catarama at the south. The reference altitude of 650 m in Santo Domingo is reduced to 220 m towards Quinindé at the northwest and 20–40 m towards the south. This structure originated from cones and slightly dissected old stretched plains. The maximum stretch of this zone reaches 80 km north of Santo Domingo.

Low Alluvial plains: This monotonous relief is located south of the Central High Plains at altitudes of 20 m or lower. The zones susceptible to floods are located in the estuary of the Guayas River. The transitions to the marine environment are the mangroves that abound in the islands formed by the sediment depositions accumulated in the Guayas River. In the contact point of the plain with the western Andean range appear some coalescing cones forming the foothills. This formation covers an area of 40 km (between Bucay and El Triunfo), characterized by a gentle slope to the west. The narrow 10–15 km strip that extends to the south has similar characteristics of the area described above and is embedded between the eastern Andean foothills and the mangroves to the west.

1.4.3 Amazonia Relief

The relief of the Ecuadorian Amazonia is divided into three types of forms (Winckell, Zebrowski, et al., 1997): (a) Sub-Andean reliefs, (b) Peri-Andean piedmont, (c) Peri-Andean hills and valleys.

Sub-Andean Reliefs

This zone is formed by mountainous and sub-mountainous relieves (2500 and 500 m in altitude, respectively). Local names like Lumbaquí, Huacamayos, Cutucú, Cóndor mountain ranges indicate the highest landscapes present in the Sub-Andean reliefs (Winckell et al. 1997b).

In the central part of this formation appears an extensive piedmont, with reliefs that stand above in El Tena, which is characterized by a long, solid dome, with ravines and canyons, which circulates the hydrographic network of the Quijos River. The highest point is the Sumaco volcano with 3900 m, but most high points are not over the 2000 m in altitude.

These reliefs are separated from the eastern Andean slope by an almost continuous meridian depression, except at the far north. The slope begins at El Chaco in the north and continues through El Puyo forming a narrow corridor with steep flanks, which then extends to the south as far as Zamora. This corridor widens between Macas and Mendez forming a small depression which allows the flow of the Namangoza River.

The relief of the central part virtually forms anorifice at the exit of Pastaza River located between the end side of the Napo River to the north and the foothills of the Cutucú mountain range to the southeast, with altitudes ranging from 600 to 900 m. Slightly hilly reliefs are present in this formation as well as fossilized and burnt structures of detrital remains of the piedmont.

Peri-Andean Piedmonts

The Peri-Andean piedmonts are mainly located in the central region of the Ecuadorian Amazonia, on both sides of the Pastaza River. Two main types of formations are part of this zone: (a) High Piedmonts or Plateaus, and (b) Low Piedmonts.

The High Piedmonts starts from 1500 m of elevation west of El Puyo, dropping gently to 300 m near the border with Peru. This formation is a big old dissected tabular spread cone that stretches to the east, where it changes to elongated dissected shreds delimited by cliffs, hills, and narrow peaks that get smoothed as it moves to the east.

The Low Piedmonts are located below the Sub-Andean Reliefs, stretching from the Napo province at the north to the Cutucú mountain range to the south, and are formed by flat stepped levels, sloping to the east, with evidence of the continuous alluvial depositions of the Amazonia hydrographic system. The relief is not higher than 600 m to the east, and examples of these formations are the Shushufindi plains in the north and Cangaime plains in the south.

Peri-Andean Hills and Valleys

This monotonous relief, less than 300 m in altitude, is formed by hills developed over tertiary sedimentary rocks and continuous coatings from quaternary spreads. The monotony is intensified by the forest cover evenly overlaying the relief. Large alluvial plains are present among these hills, which are broad valleys located mainly between the

Napo and Aguarico rivers. The relief also includes swamps and flooded areas that were formed as a result of the overflow of the rivers that form the Amazonia hydrographic system.

1.4.4 Galapagos Islands Relief

The Galapagos Archipelago is a set of 5 large and 14 medium islands and 68 islets or rocks that cover an area of approximately 8000 km^2, located on the equator, 1000 km west of the Ecuadorian coast.

Morphologically, the following groups can be recognized in the archipelago: (a) Low Elevation Islands (Baltra) with gentle undulating relief and heights that do not pass 220 m a.s.l.; (b) Shield Volcanoes without caldera (Floreana, Santiago and Santa Cruz), which are high ancient volcanoes (920 m a.s.l.) with no activity, except the lava activity of Pinta and Santiago; and (c) Large Shield Volcanoes with crest calderas (Fig. 1.9) (Winckell et al. 1997b).

The following volcano structures are present in the Galapagos Islands: (a) Low volcanoes with small summit caldera like Genovesa and Pinzon (435 m), (b) Remnants of large destroyed volcanoes with an altitude close to sea level (Marchena Island, Volcano Ecuador), and (c) Majestic volcanoes like Fernandina and Isabela islands, with altitudes of 1660 m (Wolf volcano), and 1080 m (Sierra Negra Volcano). The latter are shield-shaped volcanoes, with gentle lower slopes, strong upper flanks and tops with a hole on the center formed by the caldera sinking. Some of these shields have slight fumarola activity, and flanks covered by vegetation (e.g., Alcedo volcano), and others, such as the Sierra Negra and Cerro Azul volcano, still have volcanic lava emissions.

1.5 Geodynamic Evolution

Geologically, Ecuador is located at the junction of the Central Andes with the Northern Andes, at the level of the Galapagos Rift, which is an area of oceanic spreading from east to west that separates the Nazca plate (south) from the Cocos plate (north) (Collot et al. 2009; Marocco and Winter 1997).

The Coastal Plane is formed by magmatic rocks of oceanic origin that were coupled to the mainland at the end of the Cretaceous (about 80 million years ago), covered by marine sedimentary areas of Tertiary or Quaternary age (Fig. 1.10). The Andes are formed by continental crust covered by deformed sedimentary and igneous rocks of different ages. They were formed as a result of a complex geological process initiated in the Precambrian, in which alternated cycles of sedimentation, magmatism, and deformation occurred. The Western Cordillera, the youngest, is the result volcano sinking. The Eastern Cordillera was formed in the Mesozoic by sedimentary and magmatic rocks (Apsden et al. 1988; Marocco and Winter 1997). The Andean basin, located between the two cordilleras, was

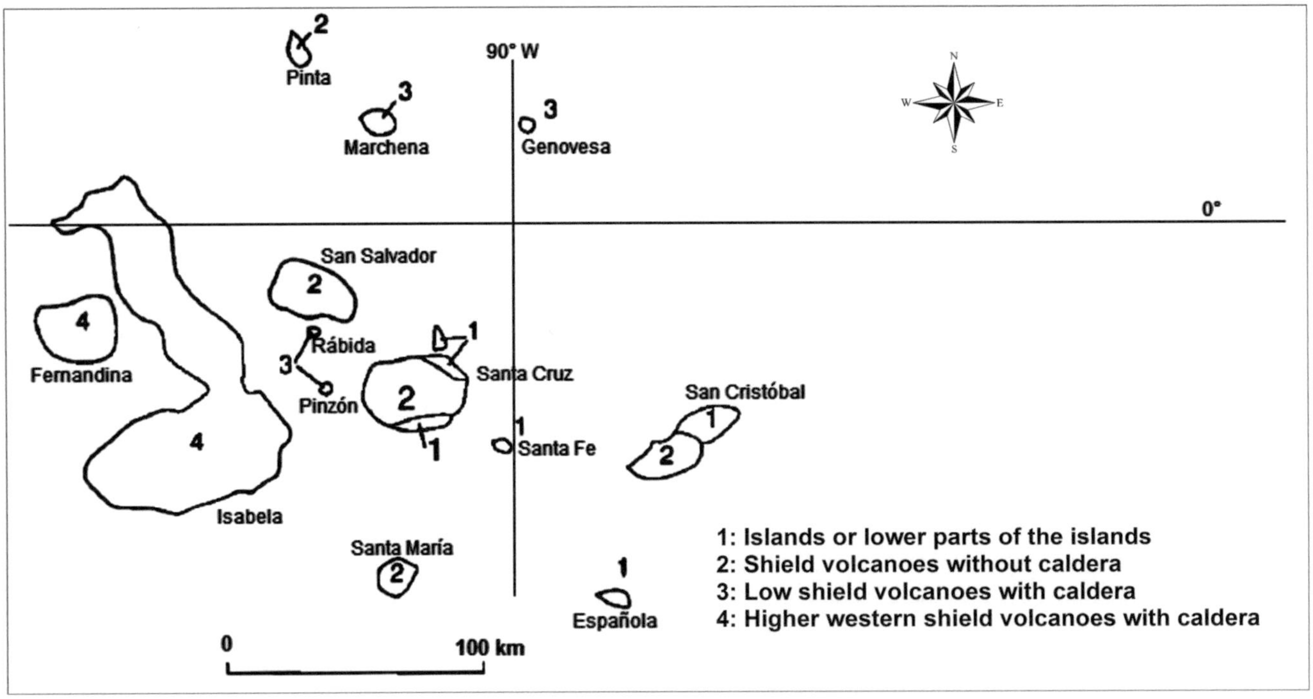

Fig. 1.9 Island types of Galapagos Archipelago (Winckell et al. 1997b)

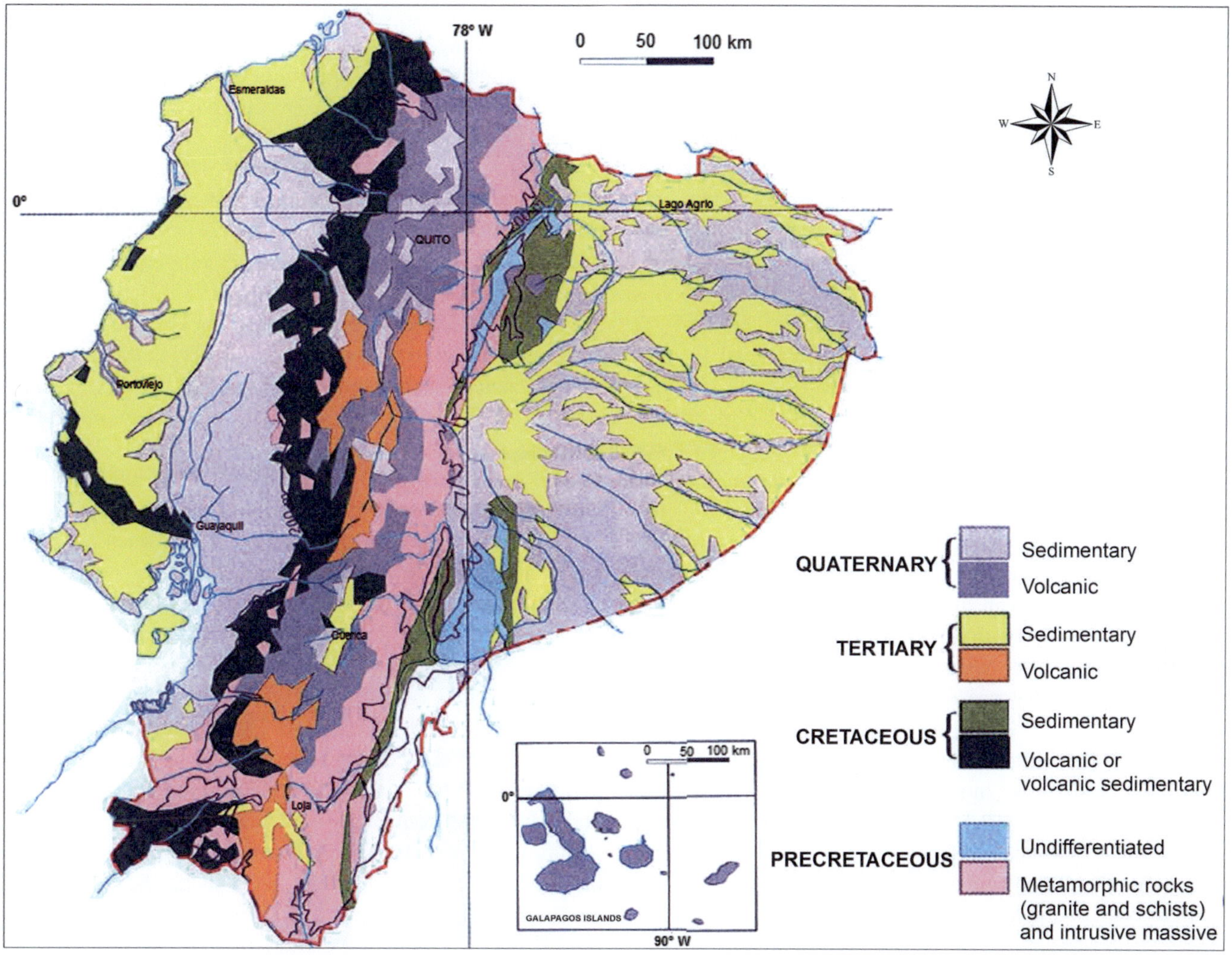

Fig. 1.10 Geological map of Ecuador (León 2010)

shaped in the Cenozoic and led to the formation of continental sedimentary valleys affected by tectonic activity and the Plio-Quaternary recent volcanism (Fig. 1.10).

The eastern foothills (sub/Andean zone), linkage of the Andes with the Guyanese-Brazilian Shield, is characterized by having low altitudes and is formed by a continental crust covered with sedimentary areas of the Paleozoic, Mesozoic and Cenozoic periods, but most of the deformations appeared during the Cenozoic (Fig. 1.10) (Marocco and Winter 1997).

1.6 Climatic Regions and Natural Vegetation

Ecuador is located in the low-pressure equatorial belt, in the influence zone of the Inter Tropical Front or Inter Tropical Convergence Zone (ITCZ), which generates relatively cool dry continental air masses over most of the country from May–June to September–October, and masses of hot humid ocean air during the rest of the year. This zone of the ITCZ affects in different ways the country regions (Huttel 1997). On the Coastal Plane, this zone is more or less lasting depending on the latitude, generating a single rainy season with a peak in March. The climate of this region is also influenced by the latitudinal oscillations of two opposing currents that are associated with masses of ocean air that accompany the displacement of the ITCZ on both sides of the equator: a mass of warm, moist air associated with El Niño warm stream coming from the Panama Gulf (north) which reaches the shores of Ecuador from December to May. The warm waters then become rains that gradually decrease toward the south. The other marine stream is the Humboldt cold stream, coming from the south, associated with masses of cool, dry air that prevail on the Ecuadorian coast the rest of the year. This stream generates a dry and fresh season that is most pronounced south of the country. The movement changes of these two streams probably caused by winds, results in anomalous rainfall patterns which are low when the Humboldt stream predominates over the El Niño stream, and are strong when El Niño prevails over the Humboldt, as happened in 1982 and 1983, when El Niño was perceptible in the region.

On the Highlands, the ITCZ follows the apparent sun movement, which crosses the equator twice a year, triggering rains in both sides stimulating the presence two wet periods separated by two dry periods. The first rain period (March–June) corresponds to the ITCZ ascend to the north. The second cycle of rains (October–December) corresponds to the southward descend of the ITCZ, as far as latitude 5 °S. In the dry period, masses of fresh dry continental air predominate, coming from the south to east.

On the Amazonia region, the passage of ITCZ practically does not affect the normal mass of warm, moist air, promoting a rainfall regime usually uniform throughout the year, with a peak in the months of June and July.

Ecuador's climate can be classified according to following three parameters (Huttel 1997; Pourrut 1983): (a) Annual rainfall regime, which in turn is classified into uniform climate (marked rainy or dry season), tropical climate (rainy and dry season) and equatorial climate (with two rainy seasons and two dry seasons); (b) Values of the annual rainfall (Fig. 1.11) with arid climate (annual rainfall <500 mm), dry climate (500–1000 mm), humid (1000–2000 mm) and very humid climate (>2000 mm); and (c) Mean annual temperature range which includes megathermic climates (annual average temperatures exceeding 22 °C), mesothermic climates (12–22 °C), and cold climates or microthermic (temperatures below 12 °C). These parameters, in turn, divide the country into nine zones of climatic influence (Fig. 1.12). On the Coastal Plane: (a) Megathermic arid tropical climate, (b) Megathermic dry tropical climate, and (c) Megathermic humid tropical climate. In the Highlands: (a) Mesothermic humid equatorial climate, (b) Mesothermic dry equatorial climate, (c) Cold equatorial climate, and (d) Megathermic very humid equatorial climate. In the Amazonia climate is classified as megathermic uniform very humid. In the Galapagos Islands it is classified as insular equatorial climate.

With regard to the vegetation, the main factors that determine the distribution of natural vegetation and crop vegetation are temperature and rainfall. The temperature in Ecuador is practically constant for each altitudinal level throughout the entire year (data from weather stations show that there is a decrease in average annual temperature of 1 °C per every 300 m of altitude) (Huttel et al. 1999). In addition, other weather parameters also affect crop distribution; this way, in the Coastal Plane, high insolation values are present only along the coastline, and then they rapidly decrease inland due to fog during the dry season (drizzle) on the interior reliefs or due to the almost constant clouds of the Andean Piedmont. These conditions generate a positive effect by prolonging the wet season beyond the end of rainy season, allowing the extension of certain crops outside their optimal climatic zone. The inter-Andean basins, and to a lesser extent the Amazonia basin, have high values of insolation (Fig. 1.13).

The chronological review made by Huttel (1997) on the classification of natural vegetation cover of Ecuador mentions the work by Acosta Solís (1977), who proposed a geo-botanical classification based on temperature and precipitation, and the work by Cañadas (1983), who combined the two climatic factors according to the Holdridge method, generating bio-climate cartography specifying the location of several characteristic plants of the country. However, Huttel considered that both of these classifications were not perfect and that the classification proposed by Naranjo (1981) is more comprehensive. This last classification, along with the information proposed by Patzelt and Echeverría (1985) and by Sarmiento (1987), provide complete information on the types of vegetation prevalent in Ecuador.

Huttel also made his own observations during field trips (collected over 2000 herbarium specimens) and described in detail the major groups and plant species at the different altitudinal levels as follows (Figs. 1.14 and 1.15): (a) Low altitude level (upper limit around 1800 m at the northern and central part of the country, and 1600 m at the south, temperatures >20 °C), with characteristic vegetation for arid, very dry, dry, wet, very wet, and hyper-humid zones, including azonal formations and the Galapagos Islands, (b) Intermediate level (1800–3200 m in the North, and 1600–2800 m in the South, average annual temperatures 13–20 °C), including the vegetation from outer slopes of the Andes, the intra-Andean depressions and azonal vegetation, (c) High floor (lower limit of 3200 at the north and 2800 at the South, and upper limit around 3600 m), with hyper-moist, very humid, humid and dry vegetation zones; and (d) Very high level (average annual temperature <10 °C), including shrubbery vegetation and the páramos.

1.7 Hydrology

The Hydrology depends on the relief, the degree of rock alteration, climate and vegetation (Pourrut 1983), topography and rainfall being the most important factors influencing the shape and density of the hydrologic network. Therefore, rivers, topography, and rains closely interact in a strong correlation. Ecuador, with its peculiar relief, has a great diversity of hydrological regimes and significantly heterogeneous surface water systems related to the diversity of physical-climate conditions.

These conditions have advantages and disadvantages. One advantage is the abundance of water resources for the country. There is a total mean runoff of 432,000 hm^3/year

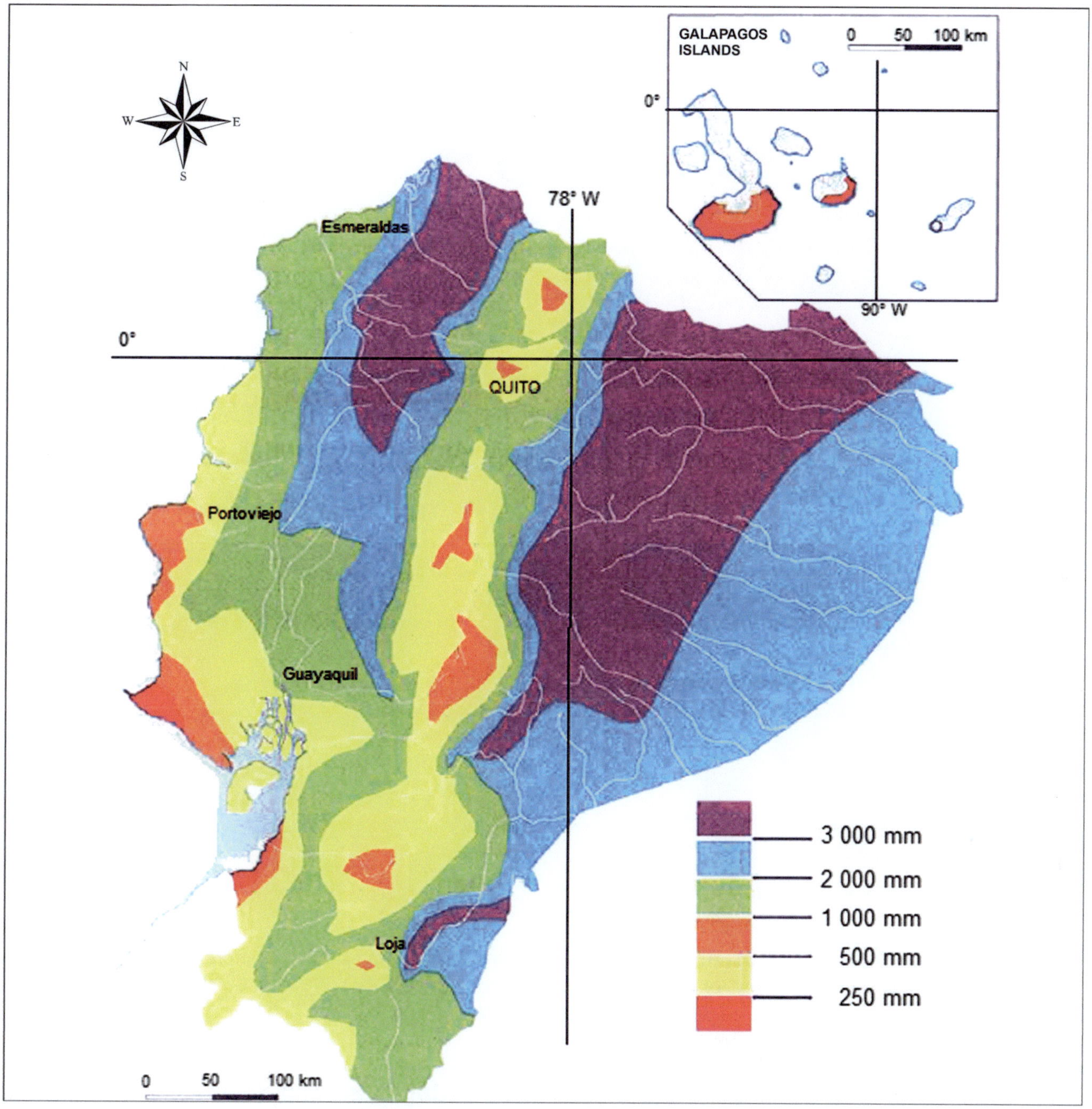

Fig. 1.11 Annual rain pattern in Ecuador (León 2010; Pourrut 1983)

(volume of water from precipitation that runs off in surface and underground rivers), with a specific runoff of 1600 mm/year, well above the 300 mm/year world average (FLACSO et al. 2008). Taking into account the temporal and spatial irregularities, the mean water availability rate is 14,809 m^3/capita/year, higher than the world average of 10,800 m^3. For this reason, the hydroelectric gross potential of Ecuador is around 643,000 GWh/year (gigawatt hours/year), at the moment only 6200 GWh/year are used (León 2010). The main disadvantage is the active soil erosion process promoted by excess runoff. In the Highlands, this specific type of soil degradation removes an average 1000 t of soil/km^2/year which, in some cases, it can increase to amounts up to 6000 t/km^2/year (Pourrut 1983).

The hydrological classification of Ecuador was made relying on the regimes used in river classification, which is based on the nature and origin of high water (rain, snow, glaciers), and the simplicity or complexity of the annual flow

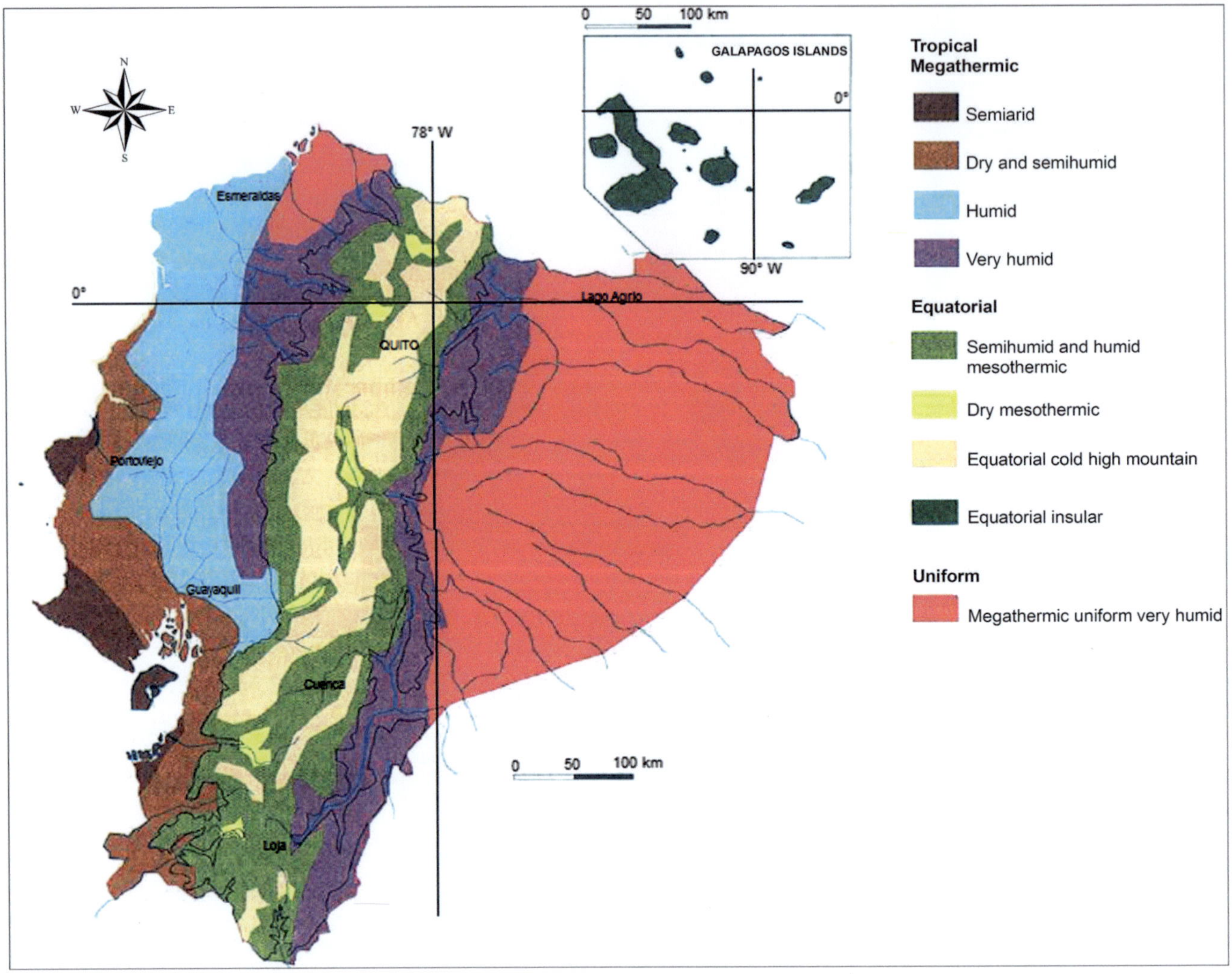

Fig. 1.12 Climate types of in Ecuador (León 2010; Pourrut 1983)

variation (single, mixed, and complex). According to this classification, the nine major classes of networks and hydrological regimes of Ecuador are as follows (Pourrut 1983): (a) Semi-humid tropical regime from the Coast, with annual rainfall ranging from 500 to 2000 mm, (b) Very humid tropical regime from the northern Coast, rainfall exceeding 3000 mm, (c) Arid tropical regime from the southern Costa, <500 mm of rainfall, (d) Complex tropical regime from the Guayas Basin, (e) Complex tropical regime from the surroundings of the Guayas Basin (exterior slope of the Western Cordillera and particularly the Angamarca, Prieto, Chanchán, Bulubulu, and Cañar rivers), always with a rainfall rate >2000 mm from January to April, (f) Glacial—snow fusion interandean regime, (g) Southern Andean regime with eastern influence, (h) Glacial—snow mountain regime, and (i) Persistante very wet regime from the Amazonia, with rains that vary from around 2000 mm/year (Palora and Zamora Valleys) to 6000 mm/year (area around the Reventador volcano).

1.8 Pedogenesis and Soils

The information on pedogenesis of Ecuadorian soils was published by Zebrowski and Sourdat (1997) and Huttel et al. (1999). These authors grouped the country soils into three types, based on the materials on which they developed. These group types are: (a) Alluvial soils, (b) Soils on volcanic ash, and (c) Soils on other materials (ancient rocks). Categories for their morphological, mineralogical and physical-chemical characteristics were defined within each group (Fig. 1.16 and Table 1.1).

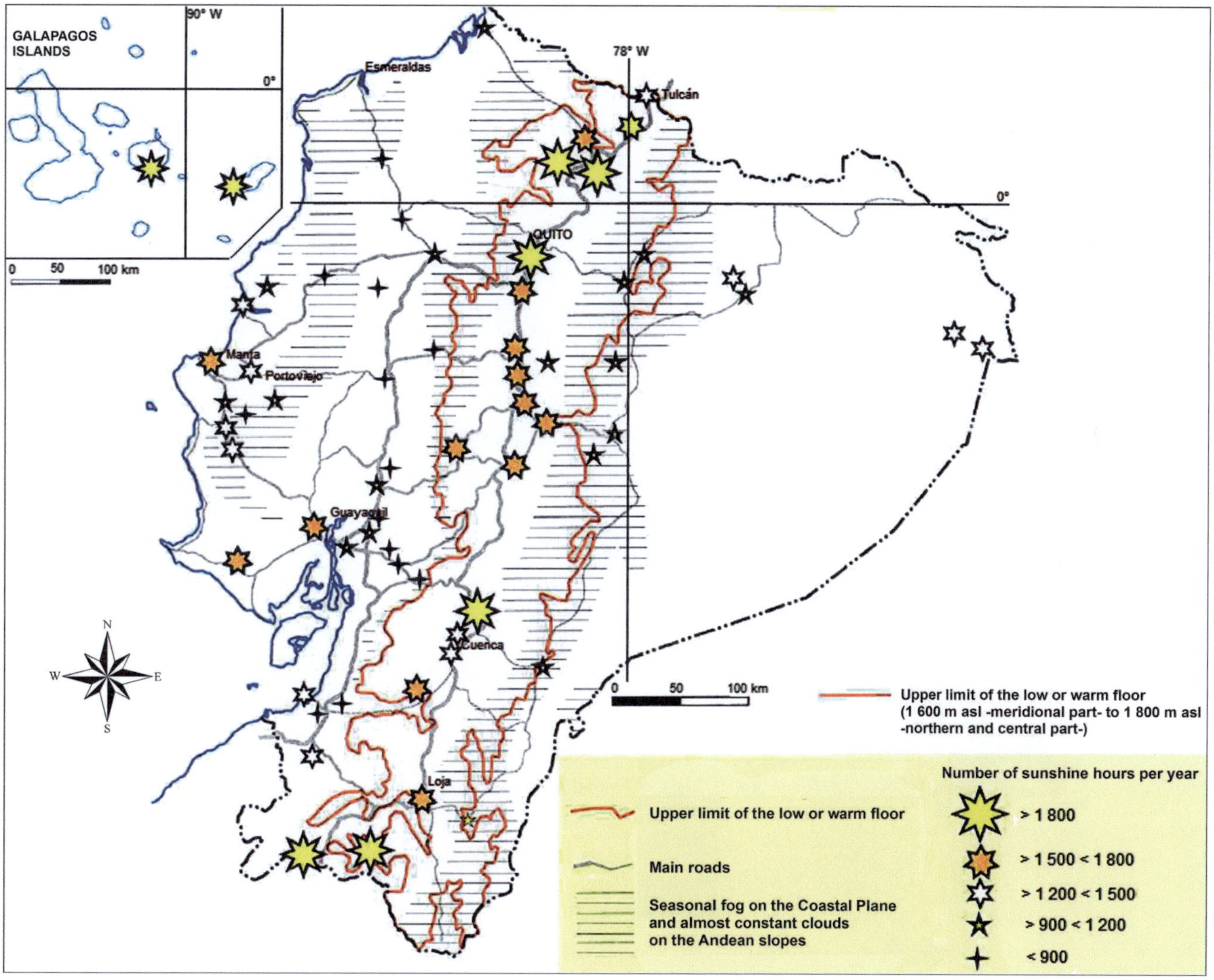

Fig. 1.13 Distribution of the main areas where fog or clouds are present (permanent or seasonal), and ranges of annual sunshine (Huttel et al. 1999)

1.8.1 Alluvial Soils

The alluvial soils were formed on recent sedimentary materials and are present in two distinct well-defined environments, known as fluvio-marine and fluvial environments.

The soils of the fluvial-marine environment are located on the beaches, coastal strands, and mangroves with loam or clay loam waterlogged and saline soils, and the soils located in the back of the mangroves, which have developed over emerged areas, without vegetation due to the high salt content, and are used to build shrimp breeding ponds.

The fluvial environment soils can be: (a) Mineral fluvial soils, mostly alluvial soils from the Coastal Plane and soils from a large section of the Amazonia. They can be divided into groups depending on the texture and the saturation degree of the exchange complex, or (b) Organic fluvial soils, located in depressed areas, mainly in the Amazonia, where the constant wetting of the profiles allows the accumulation of plant remains partially decomposed. These organic hydromorphic soils are rarely minerals and have gley spots along the profile. The organic horizon consists of fibrous materials of considerable thickness (up to 2 m), very acidic (pH $\approx$ 4.5), low base content, and desaturated (5%). The deeper mineral horizons have a loam or clayey loam texture, fluid or plastic consistency, are moderately acidic (pH 5.7), rich in bases, and saturated (50–90%).

1.8.2 Volcanic Ash Soils

Soils derived from volcanic ash occupy more than 30% of the country territory and are important from the agriculture point of view. They are very diverse, and are present from sea level up to 5000 m. The diversity of their characteristics

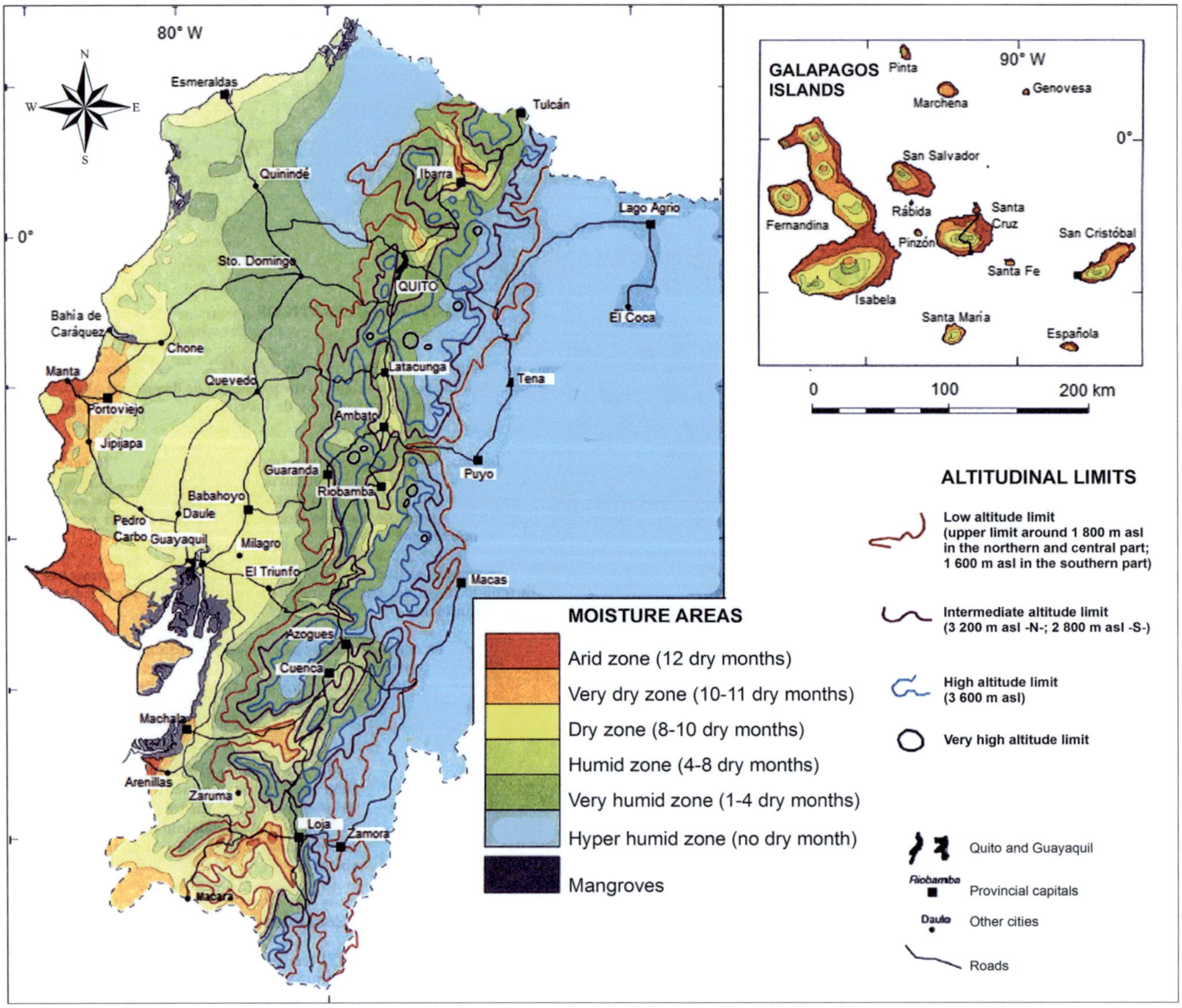

Fig. 1.14 Altitudinal levels and moisture areas (Huttel et al. 1999)

leans on the variety of climates under which they developed, and their pedogenesis depends on the nature and age of the materials from which they were formed. It is considered that all the ashes from which current soils were formed have an age between 2600 and 1600 years, calculations based on the periods of the volcanoes eruption (Zebrowski and Sourdat 1997).

Some of the volcanic soils are those developed on materials coming from volcanic emissions. In the Highlands they are presented as lapilli deposits of considerable thickness. In the higher regions of the Andes, ash-derived soils have disappeared by glacial erosion, especially from the last glaciation. In some dry regions of the inter-Andean valleys, the erosion of recent ash has uncovered the old deposits present as cemented layers known as cangahuas.

Recent ashes covered the old ashes, but, in addition, they also covered areas where they were eroded because of glaciers. The thickness of ash deposits decreases southwards as the presence of volcanoes that had activity disappears. South of Azogues, ashes can only be found in the paramos, while ashes that were deposited in drier watersheds have been lost by erosion. The soils from Zaruma are evidence of the great extent of original ashes.

In certain locations on the Coastal Plane, like Quinindé, soils developed on recent ash reach a thickness of about one m, while the lower layers formed from old ashes are approximately 10 m deep. These soils are located on slightly undulating relief. To the west, where the relief is more pronounced, soils developed from ashes have been lost by erosion and only few traces remain in protected sites. Ash

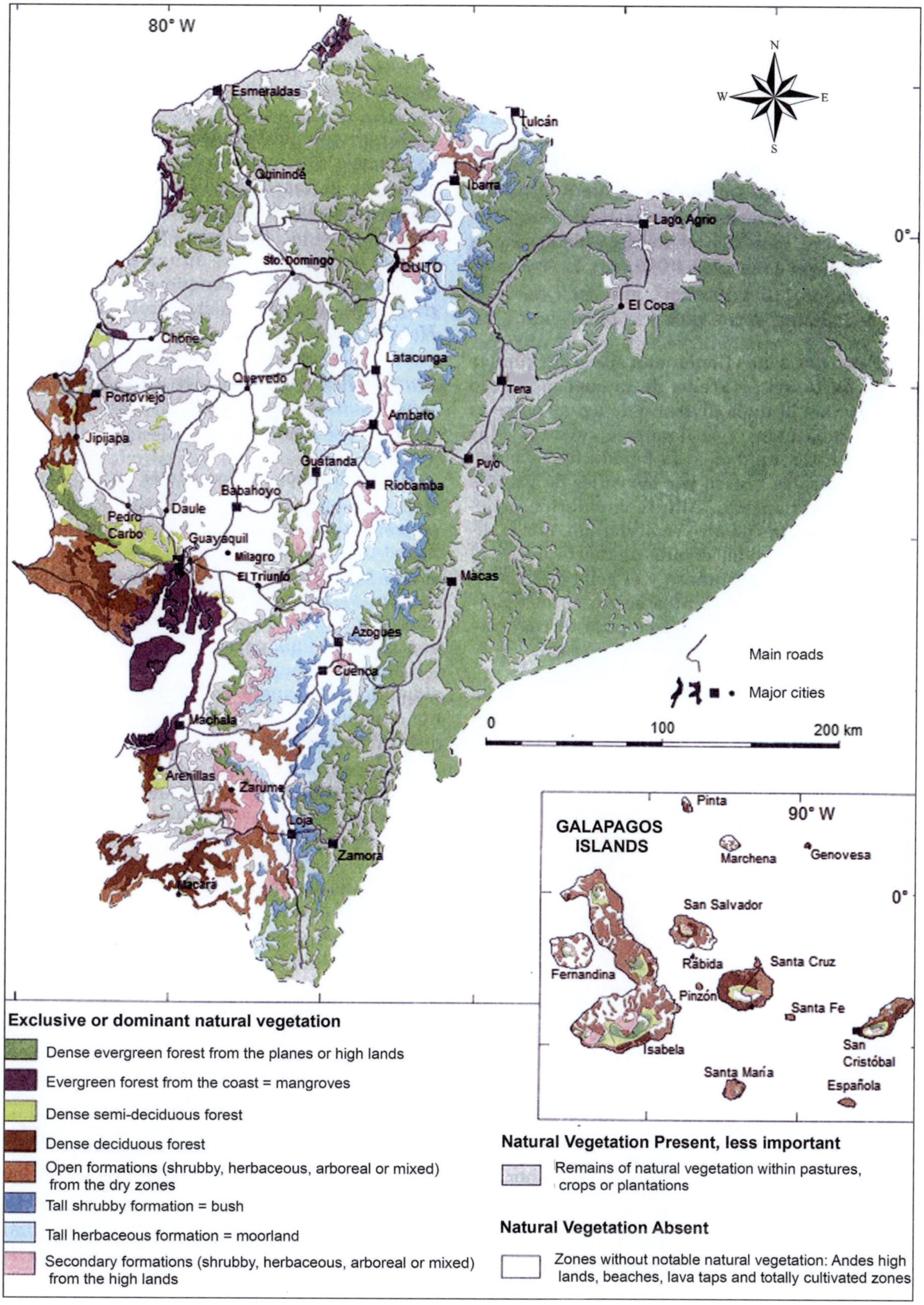

Fig. 1.15 Distribution of natural vegetation formations in Ecuador (Huttel et al. 1999)

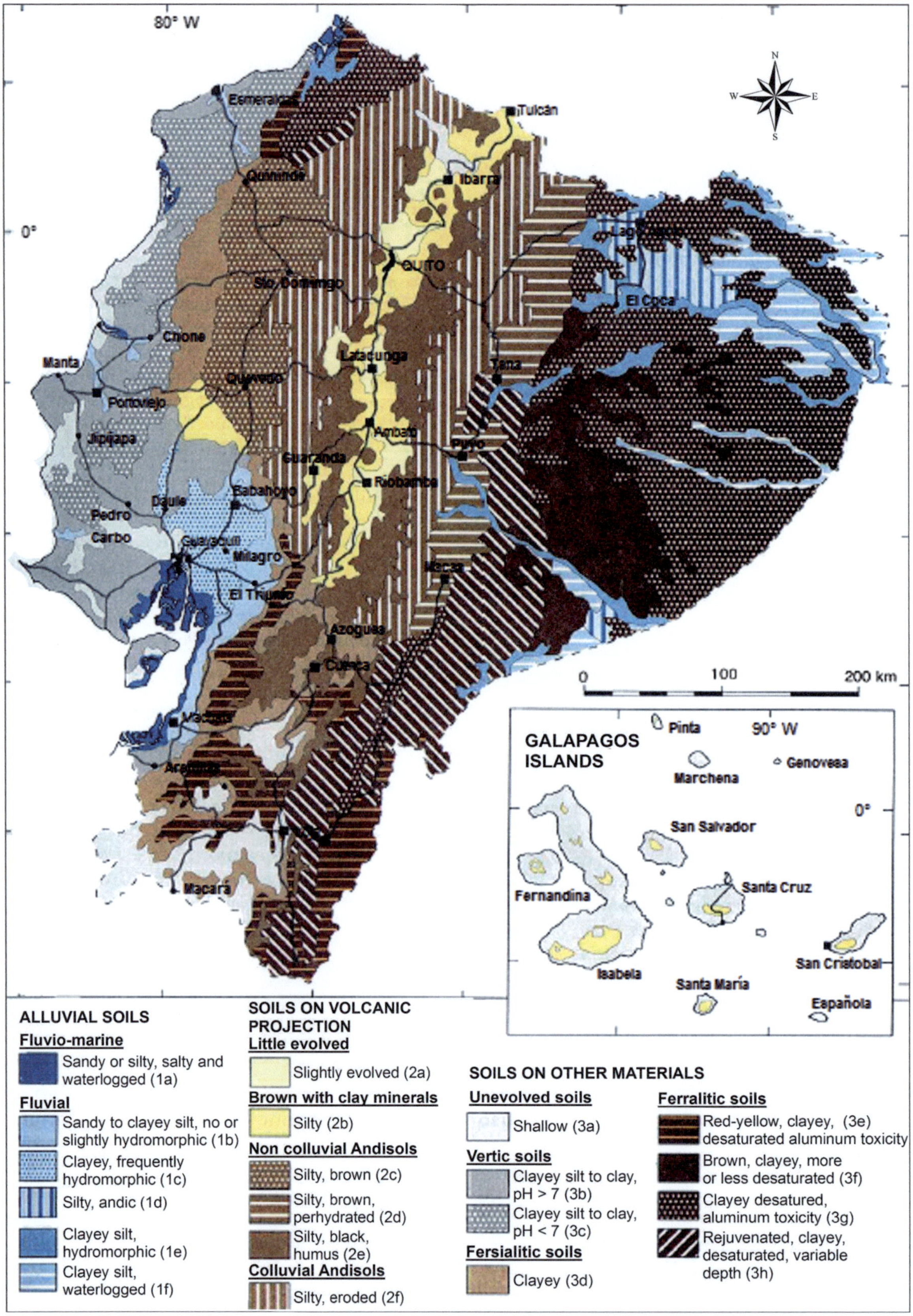

Fig. 1.16 Classification of Ecuadorian soils according to parental material (Huttel et al. 1999) (label description is included in Table 1.1)

Table 1.1 Classification of Ecuadorian soils according to parental material (Huttel et al. 1999)

Landscape	Soils	Limitations	Nomenclature
Amazonia			
Hilly	Desaturated ferralitic and pseudo-ferralitic	Low fertility, aluminum toxicity	3g
Hilly with a volcanic projection cover	Perhydrated Andisoles	Low fertility	2d
Dissected plateau	Brown ferralitic	Low fertility, localized aluminum toxicity	3f
Hilly with steep slopes	Rejuvenated ferralitic	Steep slopes, low fertility and aluminum toxicity	3e–3h
Alluvial plain	Alluvial andic	Low fertility	1d
Fluvial sediments	Hydromorphic	Water excess	1e
	Hydromorphic organic	Permanent water flooding	1f
Highlands			
Northern zone with volcanic projections			
Exterior flank	Eroded and colluvial Andisoles	Heterogeneous soils, very steep slopes	2f
Interior flanks and inter-Andean watersheds	Brown, silty	Steep slopes	2b
	Sandy	Low organic matter and water content	2a
Southern zone without volcanic projections			
Eastern exterior flanks	Rejuvenated ferralitic	Heterogeneous soils, low fertility, aluminum toxicity and very steep slopes	3h
Western outer slopes, and inter-Andean watersheds and flanks			
	Ferralitic	Low fertility, aluminum toxicity, steep slopes	3e
	Fersilalitc	Steep slopes	3d
	Vertic	Alkaline pH	3b
	Eroded poorly evolved	Shallow soils	3a
Paramos (altitude >3800 m a.s.l.)			
	Black, high humus content	Low temperature	2e
Coastal Plane			
Hills and plateau	Red and yellow ferralitic	Low fertility, aluminum toxicity, steep slopes	3e
	Rejuvenated ferralitic	Low fertility, aluminum toxicity	3h
	Ferralitic and pseudo-ferralitic	Low fertility, aluminum toxicity	3g
	Fersialitic	Steep slopes	3d
	Brown, vertic	Steep slopes	3c
	Poorly evolved	Shallow soils	3a
Gentle slope hills	Vertic	Strong pH	3b
Plains with a volcanic projection cover	Andisoils	No limitations	2c
	Brown soils	No limitations	2b
Plains with a volcanic projection cover	Vertic	Strong pH	3b
Alluvial plains	Alluvial sandy and silty	Coarse texture	1b
	Alluvial clay	Fine texture	1c
	Hydromorphic	Water excess	1e
Fluvial-marine alluvial	Salty	Stagnant water and salt	1a
Galapagos Islands			
Steep slope volcanoes	Poorly evolved	Shallow soils	3a
	Andisoils y ferralitic soils	Shallow soils	2c–3d

deposits decrease towards the South and in Balzar they only reach a thickness of 30 cm.

In the Amazonia basin, the great alteration due to severe weather makes it impossible to differentiate old from recent ashes.

1.8.3 Soils on Ancient Rocks

These types of soils cover about 60–70% of continental Ecuador and form the majority of soils present in the Peri-Andean Amazonia, the Coastal Plane and the southern third of the Highlands. Among these old rocks, sedimentary, metamorphic and igneous rocks can be found. The variety of soils developed on ancient rocks is explained by the diversity of pedogenetic conditions under different climates, regolith and topography. The climate, with variations in temperature and precipitation due to the altitude, is the main factor that promoted the diversity of the soils present in these regions of the country.

References

Acosta Solís M (1965) Los recursos naturales del Ecuador y su conservación. 1$^{era.}$ parte. Mexico D.F., Mexico

Acosta Solís M (1977) Ecología y fitoecología. Editorial Casa de la Cultura Ecuatoriana, Quito, Ecuador

Apsden J, Litherland M, Salazar (1988) Caracterización ambiental de la cuenca del río Paute. In: Estudio de Impacto Ambiental Definitivo Ex - post de la Subestación Eléctrica Cuenca. INECEL, Cuenca, Ecuador

Bornemiza E, Alvarado A (eds) (1974) Manejo de suelos en la América Latina. Sociedad Colombia de la Ciencia del Suelo, Cali, Colombia

Cañadas L (1983) El mapa bioclimático y ecológico del Ecuador. Editores Asociados, Quito, Ecuador

CLIRSEN (1979) Evaluación preliminar de la información recopilada sobre los recursos edafológicos. Quito: CLIRSEN: Gerencia de Operaciones. Departamento de Aplicaciones. Unidad Agricultura. Quito, Ecuador

Collot J-Y, Michaud F, Alvarado A, Marcaillou B, Sosson M, Ratzov G, Migeon S, Calahorrano A, Pazmino A (2009) Visión general de la morfología submarina del margen convergente de Ecuador-Sur de Colombia: implicaciones sobre la transferencia de masa y la edad de la subducción de la Cordillera de Carnegie. In: Comisión Nacional del Derecho del Mar (CNDM) (ed) Geología y geofísica marina y terrestre del Ecuador desde la costa continental hasta las islas Galápagos. CNDM-IRD-INOCAR, Guayaquil, Ecuador

Dahlgren R, Saigusa M, Ugolini F (2004) The nature, properties and management of volcanic soils. In: Advances in agronomy, vol 82. Academic Press, Washington

FLACSO, MAE, PNUMA (2008) Geo Ecuador 2008: Informe sobre el estado del medio ambiente. FLACSO - MAE - PNUMA, Quito, Ecuador

Gardi C, Angelini M, Barceló S, Comerma J, Cruz Gaistardo C, Encina Rojas A, Jones A, Krasilnikov P, Mendonça Santos Brefin M, Montanarella L, Muñiz Ugarte O, Schad P, Vara Rodríguez M, Vargas R (eds) (2014) Atlas de Suelos de América Latina y El Caribe. Oficina de Publicaciones de la Unión Europea, Luxembourg

González A (2015) Comunicación personal

González A, Maldonado F, Mejía L (1986) Memoria Explicativa Mapa Suelos del Ecuador. Sociedad Ecuatoriana de la Ciencia del Suelo, Quito, Ecuador

Huttel C (1997) Las grandes regiones climáticas y sus formaciones vegetales naturales. In: Winckell A, Marocco R, Winter T et al (eds) Paisajes naturales del Ecuador: las condiciones generales del medio natural. CEDIG - IPGH - IRD - IGM, Quito

Huttel C, Zebronski C, Gondard P, Bourliaud J (1999) Paisajes agrarios de Ecuador., vol V. Geografía Básica del Ecuador. Instituto Panamericano de Geografía e Historia, Quito, Ecuador

IEE (2014) Geoportal Instituto Espacial Ecuatoriano. IEE. http://www.institutoespacial.gob.ec/geoportal/. Accessed 23 enero 2016

IGM (2013) Atlas geográfico de la República del Ecuador. IGM - SENPLADES, Quito, Ecuador

León J (2010) Manual de geografia del Ecuador. Universidad Andina Simón Bolívar, sede Ecuador - Corporación Editora Nacional, Quito Ecuador

Marocco R, Winter T (1997) Bosquejo de la evolución geodinámica del Ecuador. In: Winckell A, Zebronski C, Sourdat M (eds) Las condiciones del medio natural. Centro Ecuatoriano de Investigación Geográfica (CEDIG) - IPGH (Sección Ecuador) - IRD (Francia) - IGM, Quito, Ecuador

Mejía L (1997) Suelos del Ecuador: Reconocimento general en base a su capacidad-fertilidad y mapa general de clasificación por Capacidad-Fertilidad de los suelos del Ecuador: Una interpretación básica sobre las características de los suelos del Ecuador Fundación Forestal Juan Manuel Durini; Instituto de la Potasa y el Fósforo (INPOFOS) - Instituto Geográfico Militar (IGM) - Instituto panamericano de Geografía e Historia (IPGH) - Sociedad Ecuatoriana de la Ciencia del Suelo (SECS), Quito, Ecuador

MIDENA, IEE, SENPLADES, MAGAP (2013) Memoria técnica geopedológica del cantón San Pedro de Huaca. Proyecto: Generación de geoinformación para la gestión del territorio a nivel nacional escala 1: 25 000. IEE, Quito, Ecuador

Naranjo P (1981) El clima del Ecuador Editorial Casa de la Cultura Ecuatoriana, Quito, Ecuador

Ortiz S, Loayza V, Albán G (2010) Gestión de datos edáficos en la cuenca baja del río Guayas con la implementación de un Sistema de Administración de Geoinformación -SAG. Paper presented at the XII Congreso Ecuatoriano de la Ciencia del Suelo: El suelo como factor de mitigación del cambio climático, Santo Domingo, Ecuador

Pacheco R (2009) El Ecuador: Recursos naturales agrícolas y del medio ambiente. Imprenta Colón, Quito

Patzelt E, Echeverría M (1985) Flora del Ecuador. Banco Central del Ecuador, Quito, Ecuador

Portais M (1997) Un ejemplo de geografía aplicada: el inventario de los recursos naturales renovables y la regionalización de la planificación agrícola en Ecuador. Revista Geográfica 9:19–25

Pourrut P (1983) Climas del Ecuador. Centro Ecuatoriano de Investigaciones Geográficas

Sarmiento G (1987) Los principales gradientes ecoclimáticos en los Andes tropicales. Paper presented at the IV Congr. Latinoamer. Bot, Medellín, Colombia

Shoji S, Nanzyo M, Dahlgren R (1993) Volcanic ash soils: genesis, properties and utilization. Elsevier Science Publishers, Amsterdam, The Netherlands

Soil Survey Staff (1975) A basic system of soil classification for making and interpreting soil surveys. In: Agricultural handbook N° 436. Soil Conservation Service, US Department of Agriculture, Washington

Soil Survey Staff (1990) Keys to soil taxonomy. In: SMSS technical monograph, 4th edn., vol 6. Virginia Polytechnic Institute, Blacksburg, Virginia

Winckell A, Marocco R, Winter T, Huttel C, Pourrut P, Zebrowski C, Sourdat M (eds) (1997a) Los paisajes naturales del Ecuador, vol 1 Las condiciones generales del medio natural. Centro Ecuatoriano de Investigación Geográfica (CEDIG) - IPGH (Sección Ecuador) - ORSTOM (Francia) - IGM, Quito, Ecuador

Winckell A, Zebrowski C, Sourdat M (1997b) Los paisajes naturales del Ecuador, vol IV Las regiones y paisajes del Ecuador. Centro Ecuatoriano de Investigación Geográfica (CEDIG) - IPGH (Sección Ecuador) - IRD (Francia) - IGM, Quito, Ecuador

Zebrowski C, Quantin P, Trujillo G (eds) (1997) Suelos volcánicos endurecidos. III Simposio Internacional Unión Europea - ORSTOM - PUCE - UCE, Quito, Ecuador

Zebrowski C, Sourdat M (1997) Los factores de la pedogénesis y los suelos en Ecuador. In: Winckell A, Marocco R, Winter T et al (eds) Los paisajes naturales del Ecuador, vol Condiciones generales del medio natural. Centro Ecuatoriano de Investigación Geográfica (CEDIG) - IPGH (Sección Ecuador) - ORSTOM (Francia) - IGM, Quito, Ecuador

Soil from the Coastal Plane

2

Julio Moreno, Gustavo Sevillano, Omar Valverde, Verónica Loayza, Renato Haro, and Johanna Zambrano

Abstract

The coastal plain is a region with high agricultural potential, since 43% of the land is suitable for production of a wide range of tropical crops such as cocoa, coffee, banana, oil palm, sugarcane, rice, corn, soybean, cassava, and numerous tropical fruits. The existence of a very fertile plain between Quinindé and the border with Peru permits the development of a highly profitable agriculture. Within this great plain are located the only Ecuadorian soils without limitations for agriculture; soils located on the old plain with volcanic protrusions, Andisols in the humid zone and Mollisols in the drier zone, occupying 13 million ha. There are several landscape units at the Ecuadorian coastal plain characterized for having similar climate and parent material conditions over which different soil types have developed contributing to the great diversity of the country soils. The main factor of soil formation on the Ecuadorian coastal plain, in addition to geological and topographical factors, was climate, especially precipitation. In general, there exist three broad sets of conditions grouping the diversity of soils from this region. (a) Soils developed from old rocks (sedimentary, metamorphic, or igneous) comprising the majority of soils in the western part of the coastal plain, particularly those developed from sedimentary rocks; (b) Soils from the old coastal plain which received volcanic ash depositions, material over which soils of high agronomic interest were developed; and (c) Alluvial soils developed over recent sedimentary materials grouped into soils from the fluvial-marine environment located on the beaches, littoral stripes and mangroves, and soils from the river surroundings corresponding to almost all of the alluvial soils of the region.

Keywords

Coastal plain • Soils over old rocks • Soils over volcanic ash • Alluvial soils

2.1 Introduction

The coastal plain, with the exception of the extremely humid northern part, is a region with high agricultural potential. While the steep slopes of the region mountain ranges are an undeniable limitation for crop production, the existence between Quinindé and the border with Peru of a very fertile plain enables the development of highly profitable agriculture. It is important to note that within this great plain are located the only Ecuadorian soils without limitations for agriculture; soils sited on the old plain with volcanic protrusions, Andisols in

J. Moreno (✉) · G. Sevillano · R. Haro · J. Zambrano
Instituto Espacial Ecuatoriano, Quito, Ecuador
e-mail: julio.moreno@institutoespacial.gob.ec

G. Sevillano
e-mail: gusfersv@gmail.com

R. Haro
e-mail: renoharobull@gmail.com

J. Zambrano
e-mail: johana.zambrano@institutoespacial.gob.ec

O. Valverde · V. Loayza
Ministerio de Agricultura y Ganadería, Quito, Ecuador
e-mail: amarobvalarias82@hotmail.com

V. Loayza
e-mail: veronica_loayza@yahoo.es

J. Espinosa et al. (eds.), *The Soils of Ecuador*, World Soils Book Series,
https://doi.org/10.1007/978-3-319-25319-0_2

the humid zone and Mollisols in drier zone, occupying 13 million ha (20% of the region's area) (Huttel et al. 1999).

The agricultural aptitude map, scale 1:200,000, generated by the PRONAREG-ORSTOM in 1982, shows that 43% of the coastal plain land (2.8 million ha) is suitable for agricultural use. This zone is located on flat to slightly hilly areas, with favorable conditions for a wide range of tropical crops such as cocoa, coffee, banana, oil palm, sugarcane, rice, corn, soybean, cassava, and numerous tropical fruits. Twenty-two percent of the land in the region (approximately 1.4 million ha) has hilly and dissected reliefs, which makes them marginal for agricultural use, but can be used for livestock production based on cultivated pastures. Finally, 35% of the region land (2.4 million ha) corresponds to areas with steep slopes and other morphopedologic limitations, and they can only be used for forest production and/or protection (Pacheco 2009; PRONAREG-ORSTOM 1982).

There are several landscape units at the Ecuadorian coastal plain characterized for having similar climate and parent material conditions over which different soil types have developed contributing to the great diversity of the country soils (Fig. 2.1). The main factor of soil formation on the Ecuadorian coastal plain, in addition to geological and topographical factors, is climate, especially precipitation, as the temperature tends to be isohyperthermic in the entire region. In general, there exist three broad sets of conditions grouping the diversity of soils from this region.

2.1.1 Soils Developed over Old Rocks

The soils developed from old rocks (sedimentary, metamorphic, or igneous) constitute the majority of soils in the western part of the coastal plain, particularly those developed from sedimentary rocks. The variety of soils formed on these old rocks is explained by the different pedogenic conditions like climate, bedrock, and relief. From these conditions, precipitation was perhaps the most important factor in the development of these soils, for example, pH, CEC, and base saturation decrease as precipitation increases in areas of soils developed from similar sedimentary rocks, and located in comparable topographic positions (Zebrowski and Sourdat 1997). The following groups can be identified within the soils developed from old rocks (Huttel et al. 1999): (a) Shallow poorly evolved soils located in areas with steep slopes, for example, mountain ranges where erosion prevents the formation of deeper soils; (b) Vertic soils with pH > 7, characteristic of dry arid zones from the coastal plain (for example, Santa Elena Peninsula), developed mainly from old sedimentary rocks such as clays and sandstones; (c) Brown (Mollisoles) and vertic (Vertisols) soils with pH < 7, located in the humid part of Chongón Colonche mountain range: the brown soils are located on steep slopes and vertic soils on less pronounced relief; and (d) Highly desaturated red ferralitic soils from the northern more humid zones of the Esmeraldas Province, with very acid pH resulting in aluminum toxicity.

2.1.2 Soils Developed over Volcanic Ash

Sections of the old coastal plain received volcanic ash depositions, material over which soils of high agronomic interest were developed. The variety of climates under which the ashes were weathered is largely responsible for the diverse characteristics of these soils. In the northern area (Quinindé), soils developed on recent ash have a thickness of about one meter, but they are on top of another 10-m layer of soils developed over old ashes. The thickness of soils developed from recent ash decreases as the sites get away from the Andes and move southward, reaching only 30 cm in the south (Balzar). The fine ash particles were pushed away by the wind to considerable distances from the centers of emission, so the weathering was rapid, resulting in homogeneous, loamy, and brown profiles. Rainfall, especially the length of the dry season, played a major role on developing some of the characteristics of these soils, for example, in the more humid areas (<4 dry months per year), pedogenesis lead to the formation of amorphous clays (allophane and imogolite), whereas in regions with a well-marked dry season (>4 months per year) the synthesis of crystalline clay minerals (2:1 clays) was the dominant pedogenic process (Huttel et al. 1999; Zebrowski and Sourdat 1997).

2.1.3 Alluvial Soils Developed over Recent Sedimentary Materials

These are alluvial soils formed on recent sedimentary materials that are grouped into (a) soils from the fluvial-marine environment located on the beaches, littoral stripes, and mangroves; and (b) soils from the river environment corresponding to almost all of the alluvial soils of the region. The texture and the degree of base saturation from the fluvial environment allow the differentiation of groups of generally good fertility and easy to cultivate soils, but they may have the following unfavorable characteristics: (1) Sandy texture or existence of surface stones upstream of the spreading areas; (2) Water excess and the consequent presence of hydromorphic soils making farming activity difficult or even impossible, particularly in areas where the water table rises to the surface; and (3) Clayey texture, characteristic of soils of the recent plain, that limits drainage and favors flooding in the rainy season; however, these are ideal soils for patty rice (Huttel et al. 1999; Moreno 2001; Zebrowski and Sourdat 1997).

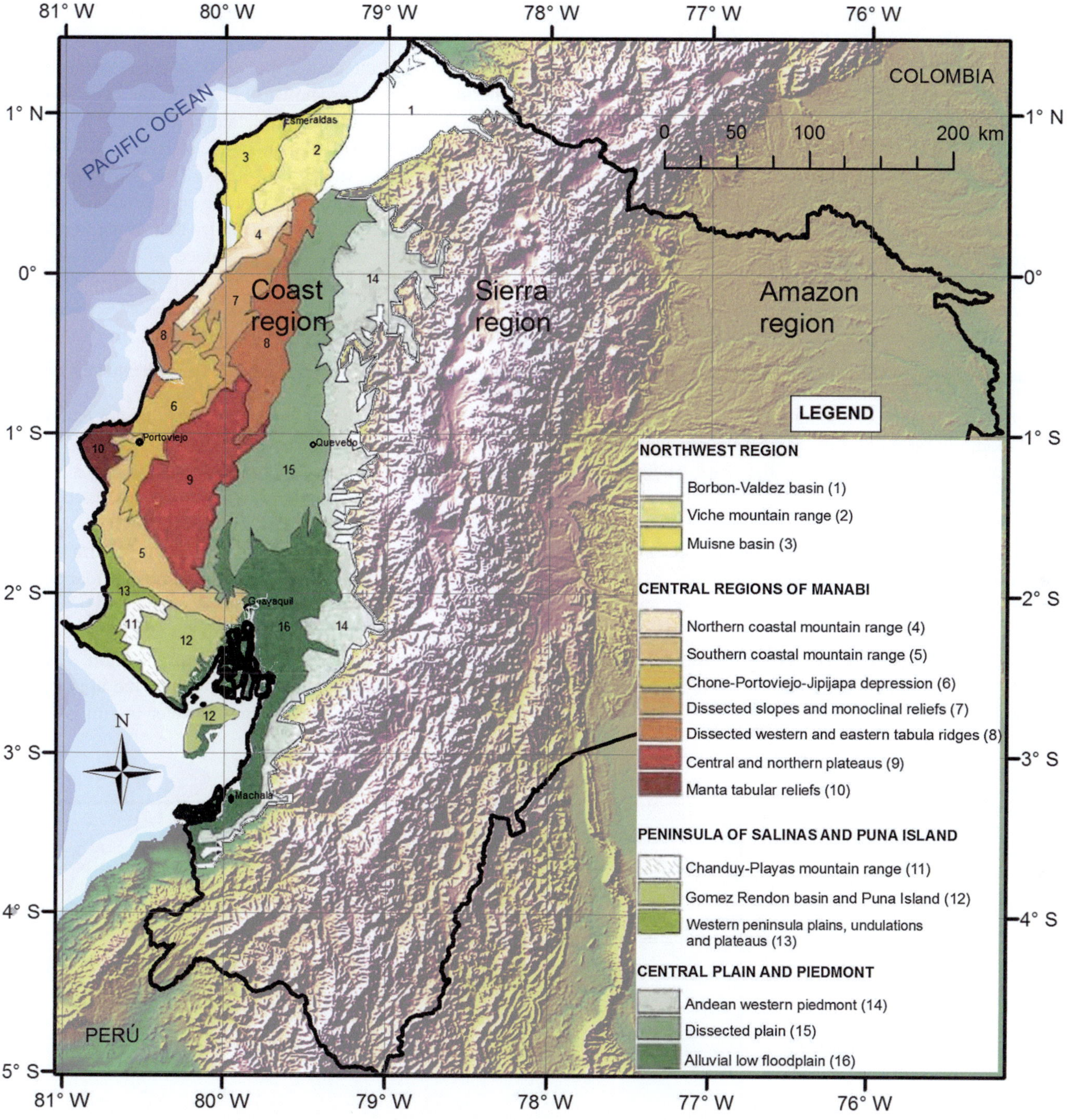

Fig. 2.1 Coastal plain landscape distribution with respect to continental Ecuador (adapted from Winckell and Zebrowski 1997)

A detailed description of each landscape unit and its respective soils of the Ecuadorian coastal plain are presented below.

2.2 Tabular Cone Totally Covered with Volcanic Ash

This area is located to the east of the Borbón Valdez basin in the province of Esmeraldas (Fig. 2.2). The connection between the Andean and coastal reliefs generally occurs through a steep slope, but in this case, the slope was attenuated in the northwestern part by the existence of a tabular relief, slightly inclined to the northwest, which ends at 750 m o.s.l. at Alto Tambo, San Lorenzo. This relief was formed from ancient coalescing spreading cones established at the mouth of the Mira, Lita, and Gualpi rivers, which have a torrential regime and drain the inter-Andean valleys (Fig. 2.2). The area consists of thick dendritic and torrential deposits (≈600 m thick), which were covered by recent volcanic projections, probably coming from volcanoes located south of Colombia. These materials, exposed to a warm and hyper humid climate (4000–6000 mm rain per year and temperature >25 °C), resulted in perhydrated volcanic soils (>100% water retention), somehow sandy at the top of the cone and silty at the bottom (Zebrowski and Sourdat 1997).

2.2.1 Areas Near the Andean Slopes

This landscape integrates the crest surfaces (b) and surrounding abrupt (c) as shown in Fig. 2.3. The crest level has a very smooth surface, moderately dissected into small rounded biconvex tops that end at the village of Alto Tambo, at the gorge outlet of the Lita River. This platform reaches heights exceeding 800 m o.s.l. at the confluence of the Mira River into the Lita River. These cones then come together to the east rapidly widening the surface of the peaks. The thickness of these perhydrated Andisols steadily decreases from southeast to northeast. Soils from the abrupt slope are eroded and colluviated, so soils of volcanic originally adjacent to shallow, clayey, and often stony soils (Winckell and Zebrowski 1997).

In this zone, there are also small areas that are remainders of steeped deposits accumulated above the cones, usually perched on the flanks of the current riversides [(a) in Fig. 2.3]. Its position into the Andean flank favored the accumulation of an important volcanic ash cover on which deep perhydrated Andisols were developed (2–3 m deep).

The representative soils of this unit are Hydrudands (Soil Survey Staff 2006), with low bulk density (<0.9 g cm^{-3}), high phosphorus fixation, and silty loam texture. These soils have a very strong reaction to sodium fluoride (NaF) throughout the entire profile, due to the high content of amorphous clays. The amount of water in situ is always

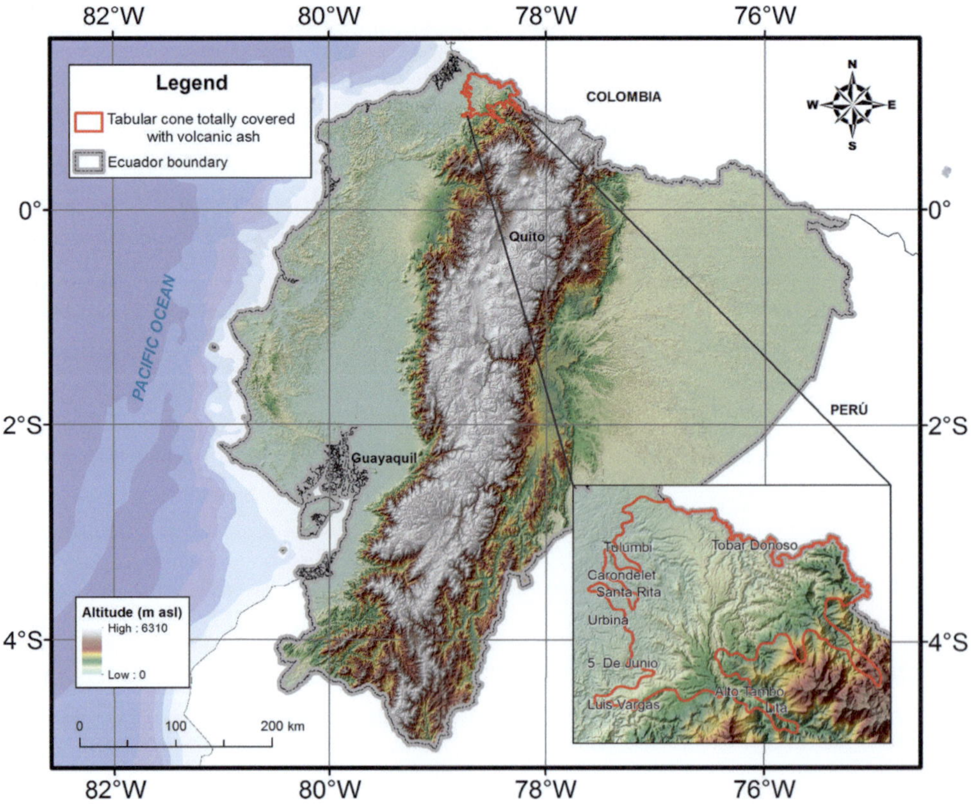

Fig. 2.2 Location with respect to continental Ecuador of the tabular cone totally covered with volcanic ash (adapted from Zebrowski and Sourdat 1997)

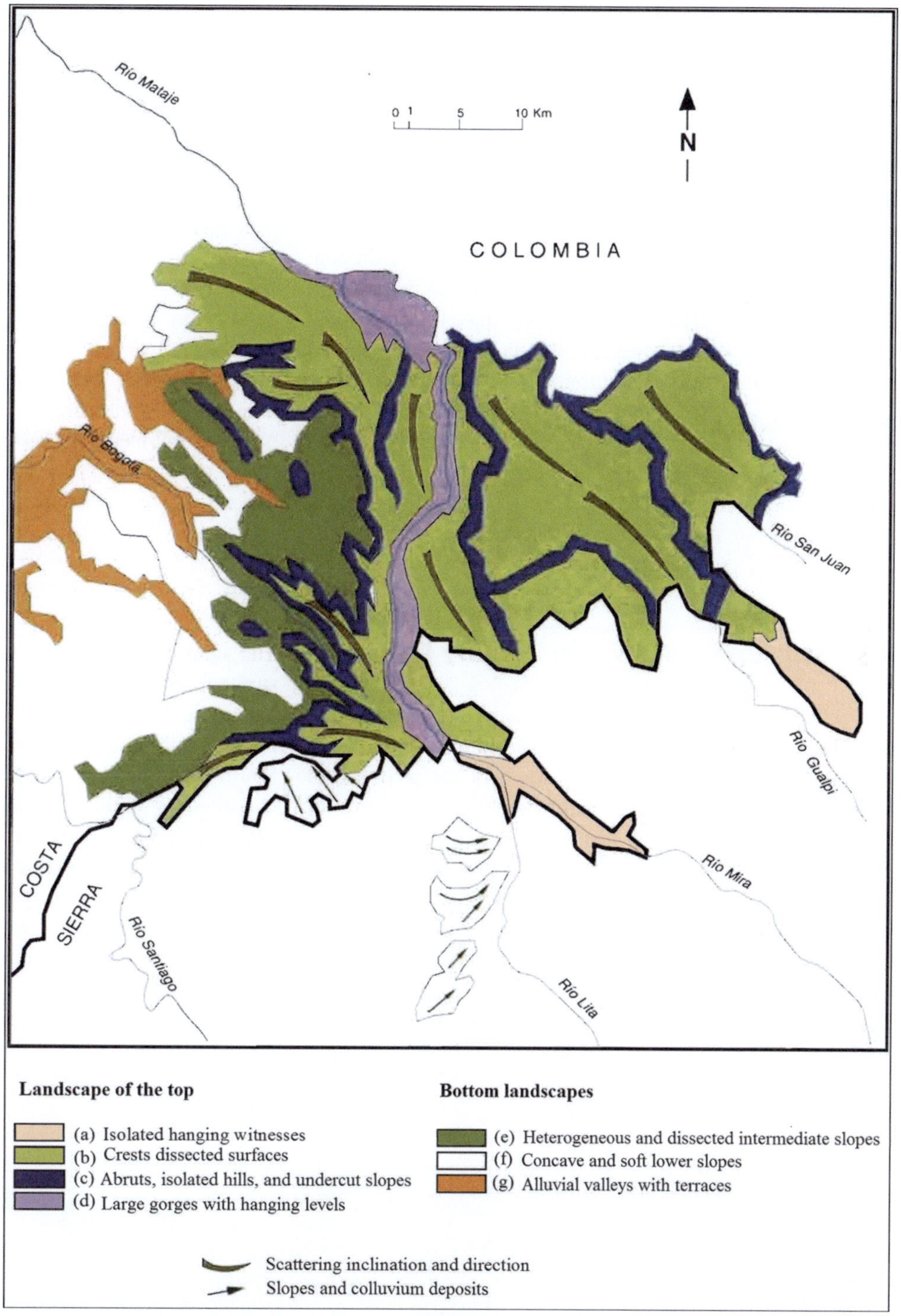

Fig. 2.3 Tabular cone of dissected relieves (adapted from Zebrowski and Sourdat 1997)

high, close to those obtained at pF3 on a wet soil, usually >200%, sometimes reaching 300%. Topsoil pH is low (4.6), but increases with depth (pH slightly lower than 6.0 in the rest of the profile). The amount of exchangeable bases is very low (≤ 0.3 cmol kg^{-1}), as well as base saturation ($\leq 1\%$) (Zebrowski and Sourdat 1997).

2.2.2 Rolling Surfaces

At the foot of abrupt, slope extends a complex set of hills and moderate slopes [(e) and (f) in Fig. 2.3)], mainly present in the western part of the cone (the most eroded), developed on relatively homogenous deposits of highly weathered sandstones and conglomerates that have formed motley clays and pebbles. These hills and slopes have less torrential facies with a state of alteration far more advanced that can only be explained by the older age of the lower deposits in relation to higher deposits, and for the more humid climate conditions. The upper part of these slopes has moderate to strong inclination (straight and convex) which are dissected by numerous parallel streams. The lower part of the slopes, or those located forward, contains concave, soft inclination slightly extended by the underlying reliefs (Winckell and Zebrowski 1997). The thin ash cover has been partially removed by erosion. Consequently, the silty perhydrated Andisols sit beside very desaturated yellow pseudoferralitic and ferralitic soils (Fig. 2.4) with clayey loam texture, low organic matter content, low CEC (<10 cmol kg^{-1}), pH strongly acid (4.6–4.1), highly desaturated and overly rich in exchangeable aluminum (Al) in a matrix of kaolinite, gibbsite traces, iron oxides, and fine quartz (Zebrowski and Sourdat 1997).

2.3 Structural Reliefs over Tertiary Deposits

The reliefs are located within the coastal plain provinces of Manabí and Esmeraldas (Fig. 2.5), but about half of this zone is located in the center and south of the province of Manabí as the characteristic landscape from this area of Ecuador. The altitude with respect to the crest tabular surfaces varies from 450 to 500 m o.s.l. in the center and from 250 to 300 m o.s.l. to the sides. All of these landscapes belong to only one type of relief developed on over an association of geological formations characterized by a hard summit layer on top of a soft bottom layer; in other words, a base of clays and siltstones (Onzole formation) topped by a set of detrital sediments of sandstones, sand, and conglomerates (Borbón formation) (Aalto and Miller Iii 1999; Ramírez 2013; Winckell and Zebrowski 1997). With regard to climate, these structural reliefs, mainly on the central and southern sections of Manabí, are characterized by receiving an average annual rainfall >1000 mm, reaching 1500 mm on the surfaces of the plateau. Average annual temperatures are >25 °C (Winckell and Zebrowski 1997).

The altitudinal distribution of the structural reliefs that is presented in Fig. 2.6 describes its components as (a) Top plateau crest surfaces with soft slopes and well-developed soils; (b) Surrounding flanks (abrupt) with steep slopes exposed to severe erosion processes and poorly developed soils; (c) Low flanks with a moderate gradient; and (d) Alluvial flat valleys with more developed soils in the higher and stable areas (middle and upper terraces) and poorly developed soils in areas close to the riverbeds (lower terraces) that are exposed to flooding originating new fluvial depositions that constantly rejuvenate these soils.

2.3.1 Sandstone Plateau

The sandstone plateau accounts for almost all of the crests surfaces. Soils from this area have mollic characters; deep epipedon, >50% base saturation and silty clay to clayey texture (Winckell and Zebrowski 1997; Zebrowski and Sourdat 1997). As an example a soil profile from a soil classified as Mollic Hapludalfs (Soil Survey Staff 2006) is presented (Fig. 2.7; Table 2.1). This soil has a 34-cm epipedon with loamy texture at the surface and clayey in the deeper horizons, the A horizon color is very dark gray and the Bt horizon is clearer with a very hard consistency which limits root growth (CLIRSEN et al. 2012a).

In addition to the plateau summit surface, a sandy abrupt contour can also be found in this unit, which sits on a cemented detrital base from the Borbón formation, surrounding the plateau and delimiting it from the adjacent

Fig. 2.4 Profile of a ferralitic soil (left) located in the area of San Lorenzo, Esmeraldas where oil palm (right) is grown

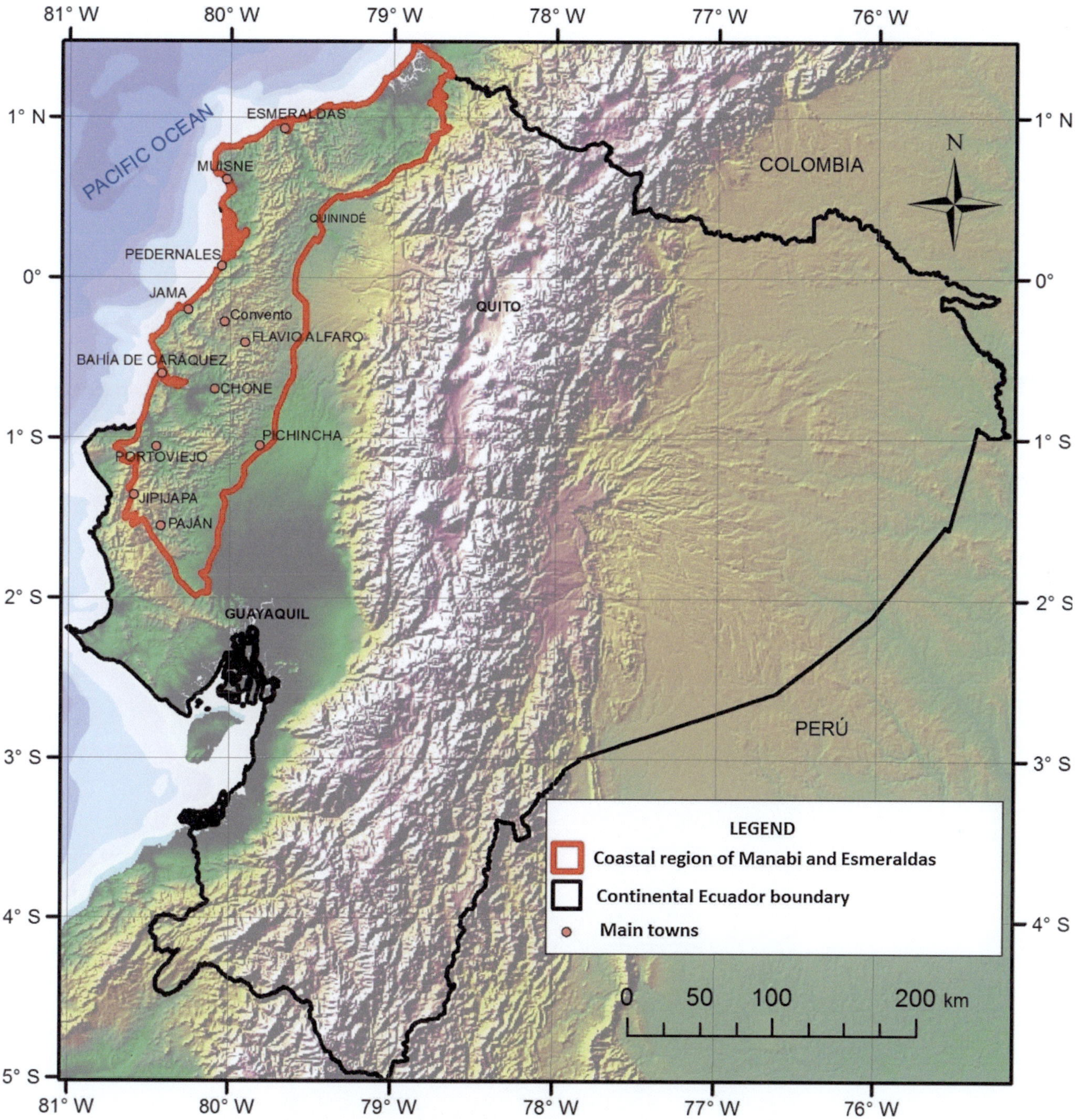

Fig. 2.5 Location of the structural reliefs over Tertiary sediments of the Manabí (central and southern) and Esmeraldas (north) Provinces with respect to continental Ecuador (adapted from Winckell and Zebrowski 1997)

areas. The altitude difference of the slopes from the higher to the lower surrounding landscapes is usually around 300 m and the inclination is almost always >70% forming rocky cliffs. Due to the steep slopes, soils at these sites are subject to severe erosion that overcomes the pedogenic processes; therefore, they are characterized for having a shallow profile, often less <10 cm, due to the outcrop of rocky material from the substrate (Zebrowski and Sourdat 1997).

A typical soil profile from this zone is presented in Fig. 2.8 and Table 2.2. This soil is classified as Lithic Ustorthents (Soil Survey Staff 2006) and its profile shows little or no evidence of pedogenic horizon formation (IEE 2015).

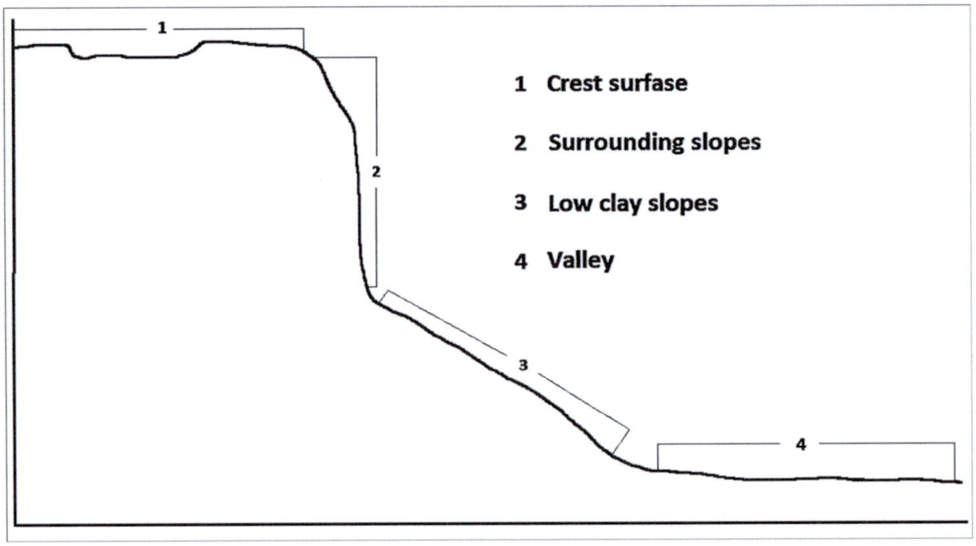

Fig. 2.6 Distribution of different landscapes developed on the structural reliefs over Tertiary sediments

Fig. 2.7 Profile of a soil classified as Mollic Hapludalfs at Peña Alta, Santa Ana—Manabí, located on the plateau crest surface from the Borbón formation (sandstones) over a <12% slope (IEE 2015)

Table 2.1 Profile description of a soil classified as Mollic Hapludalfs from the plateau crest surface (IEE 2015)

Horizons	Depth (cm)	Description
A	0–25	Very dark gray (10YR 3/1) moist color, loamy texture, granular to subangular blocky structure, pH 6.3, organic matter 2.5%, CEC 24 cmol kg^{-1}, base saturation 91.2%
Bt	25–45	Very dark gray (10YR 3/1) moist color, clayey loam texture, subangular blocky structure, clay coating, pH 6.2, organic matter 0.7%
BC	45–70	Very dark grayish brown (2.5Y 3/2) moist color, clayey loam texture, massive to subangular blocky structure with a few coarse fragments of gravel

Fig. 2.8 Profile of a soil classified as Lithic Ustorthents located at San Vicente, Manabí (IEE 2015)

Table 2.2 Profile characteristics of a soil classified as Lithic Ustorthents (IEE 2015)

Horizons	Depth (cm)	Description
O	0–5	Very dark grayish brown (10YR 3/2) moist color, loamy texture, granular to subangular blocky structure
A	5–30	Brown (10YR 4/3) moist color, sandy-clayey loam texture, subangular blocky structure, pH 7.1, organic matter 1.7%, CEC 20 cmol kg^{-1}, base saturation 97.3%
CR	30–60	Yellowish brown (10YR 5/4) moist color, sandy clay loam texture, massive structure with coarse gravel fragments showing little weathering
R	60–90+	White (2.5Y 8/1) moist color

2.3.2 Clayey Reliefs

Beyond the steep slopes extends a set of clayey slopes with a moderate gradient (<40%) moving down to the alluvial valleys mostly located at altitudes <100 m o.s.l., in places even <50 m o.s.l. These slopes have reliefs derived from the Onzole formation made of relatively homogeneous deposition of siltstones and fine poorly cemented sandstones containing minor insertions of clays and lutites. These facies, without primary coherence, and easily alterable even under dry climate, are made of soft rocks located under the hard sandstone layers (formation Borbón) from the upper tabular reliefs that rarely exceed 200 m o.s.l. (Aalto and Miller Iii 1999; Ministerio de Transporte y Obras Públicas and INECA 2012; Winckell and Zebrowski 1997).

This unit is located beyond the influence of upper orographic precipitation regime, and for this reason has a climate characterized by a long dry season which has led to moderate changes that developed heavy clayey, vertic, and saturated soils (Zebrowski and Sourdat 1997). As an example of this unit, a profile of a soil classified as Vertic Haplustepts is presented (Soil Survey Staff 2006) in Fig. 2.9 and Table 2.3. This profile shows an incipient pedogenic development and vertic characteristics, with the consequent 2-cm-wide and up to 50-cm-deep crack formation. The B horizon shows glossy slickensides on the aggregates.

Fig. 2.9 Profile of a soil classified as Vertic Haplustepts (left) located in El Hormiguero, Portoviejo–Manabí, on a plateau developed over the Onzole formation (right) (IEE 2015)

Table 2.3 Profile characteristics of a soil classified as Vertic Haplustepts (IEE 2015)

Horizons	Depth (cm)	Description
Ap	0–10	Black (10YR 2/1) moist color, clayey texture, subangular blocky structure, pH 7.1, organic matter 3.5%, CEC 42 cmol kg^{-1}, base saturation 93.9%
Bss	10–30	Very dark brown (10YR 2/2) moist color, clayey loam texture, angular and subangular blocky structure, pH 7.5, organic matter 1.3%
BCss	30–60	Very dark gray (10YR 3/1) main moist color, dark brown (10YR 3/3) secondary moist color, clayey loam texture, massive to angular blocky structure
C	60–100	Brown (10YR 4/3) main moist color, light oil brown (2.5Y 5/3) secondary mist color, sandy clay texture, unstructured

Moisture regime is Ustic, which means that the profile is dry for 90 or more cumulative days in normal years (IEE 2015).

2.3.3 Alluvial Environment

Two of the largest coastal plain valleys are situated south of the Portoviejo River and north of the Chone River in the Chone–Portoviejo depression, both valleys having high and low flat terraces. The high terraces are 6–10 m above the present river course; soils are basic, vertic, silty clay to clayey. The lower terraces, subject to annual flooding, are located 1–5 m above the water; soils are poorly evolved of sandy to silty texture (Winckell and Zebrowski 1997). Soils from the lower terraces, adjacent to the existing river channels, are constantly rejuvenated by new contributions of sedimentary material transported by the annual river flooding during rainy periods. The middle and upper terraces, also flat (<5% slope), are not exposed to new contributions from alluvial material because they are located at higher staggered levels in relation to the riverbeds and, for this reason, these terraces had the necessary stability to develop highly fertile soils, with good physical and chemical conditions, where the region agriculture is conducted. Due to river proximity, these soils have more moisture content in the dry periods or the use of irrigations is feasible.

As an example of this type of alluvial soils, the profile of a soil classified as Fluventic Eutrudepts (Soil Survey Staff 2006) is presented in Fig. 2.10 and Table 2.4. This soil is characterized for having the following horizon sequence is A/Bw/BC/C, loamy texture, good natural drainage, and a 70-cm effective depth. The Udic moisture regime indicates that the entire profile is not dry more than three cumulative months most of the years (IEE 2015).

2.4 Tabular Cone of Santo Domingo

The Great Tabular Cone of Santo Domingo (Fig. 2.11) was formed from deposits contributed by major rivers coming from northern Highlands. The amplitudes of the sediment deposition and subsequent boxing between steep banks (400 m in Mindo) are evidence of the deposits old age, which came from rivers of larger capacity and amplitude than those of the actual river network, and are related to one or various periods of glacier melting, and with later increase of fluvial

Fig. 2.10 Profile of a soil classified as Fluventic Eutrudepts (left) located at Quiroga, Bolívar, Manabí, on a middle terrace cultivated with cocoa (*Theobroma cacao* L.) (IEE 2015)

Table 2.4 Profile characteristics of the soil classified as Fluventic Eutrudepts (IEE 2015)

Horizons	Depth (cm)	Description
Ap	0–10	Very dark gray (10YR 3/1) mist color, loamy texture, subangular blocky structure, pH 7.3, organic matter 4.5%, CEC 40 cmol kg^{-1}, base saturation 97.6%
Bw	10–30	Very dark grayish brown (10YR 3/2) mist color, loamy texture, angular blocky structure, pH 7.3, organic matter 0.9%
BC	30–70	Brown (10YR 4/3) moist color, loamy texture, massive structure
C	70–100	Dark grayish brown (10YR 4/2) moist color, loamy texture, massive structure

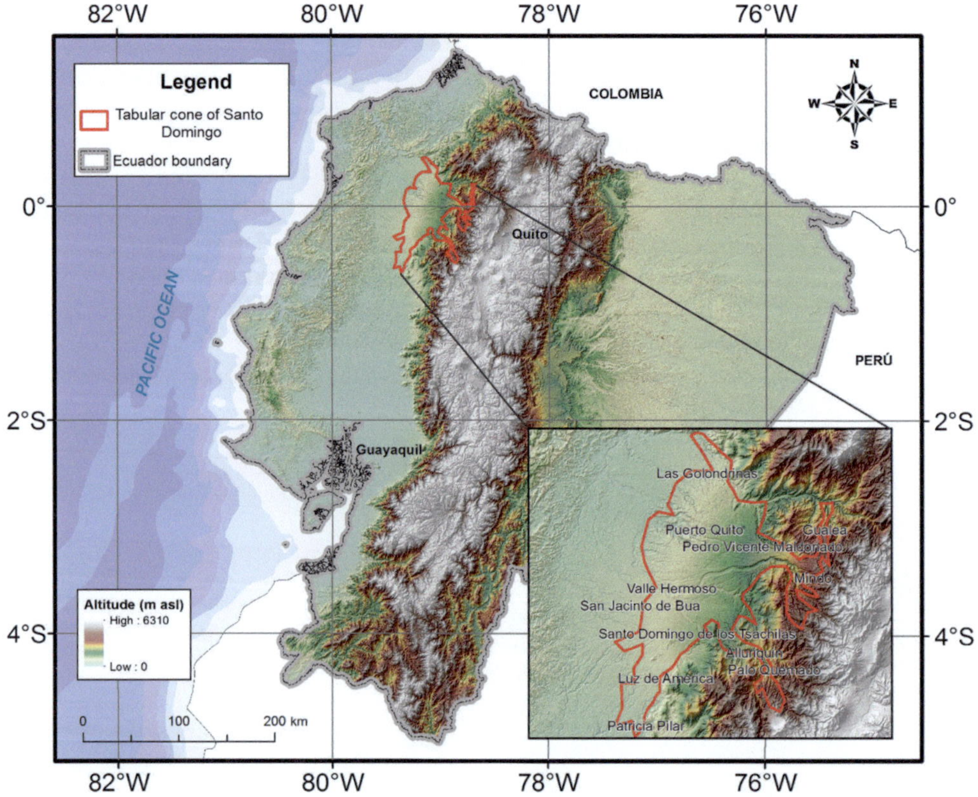

Fig. 2.11 Location of the Tabular Cone of Santo Domingo with respect to continental Ecuador (adapted from Winckell et al. 1997)

water runoff promoted by the melting of Andean glacier cap. Furthermore, it is also evident the relationship with the huge amount of deposited material (volcanic ash) product of the intense Andean volcanic activity (Winckell et al. 1997).

This large cone, situated at a humid tropical zone, receives an annual average precipitation >2500 mm, and has an annual average temperature >22 °C, generating an Udic and even Perudic moisture regime and an isohyperthermic soil temperature regime. All soils are derived from volcanic ash due to the proximity to the emission centers; ash covered all the reliefs no matter the slope. Under these climate conditions, the alteration of volcanic ash promotes the formation of low base saturation amorphous clays (allophane) (Winckell et al. 1997).

2.4.1 Areas Near the Western Andean Flanks

This zone is located at the top of the cone, along the western Andean flanks, site where the alluvial deposits have taken a perched position along the current notches formed by the boxing of the hydrographic network. These conditions have promoted the formation of an association of dissected–notched surfaces with very abrupt external flanks showing rocky conglomeratic outcrops (Winckell et al. 1997).

The modal profile of this unit corresponds to a soil classified as Thaptic Hapludands (Soil Survey Staff 2006) presented in Fig. 2.12 and Table 2.5. The profile is characterized by having very dark colors in the first horizon and strong reaction to NaF in most of the horizons.

Fig. 2.12 Profile of a soil classified as Thaptic Hapludands (left), located at La Mina, Mindo, Pichincha. The site is at the top of the cone close to the western Andes mountain range. These sites are used for silvopastoral systems to raise livestock for both milking and meet purposes (right) (IEE 2015)

Table 2.5 Profile characteristics of the soil classified as Thaptic Hapludands (IEE 2015)

Horizons	Depth (cm)	Description
Ap	0–10	Very dark grayish brown (10YR 3/2) moist color, sandy loam texture, granular to subangular blocky structure, strong reaction to NaF, pH 5.6, organic matter 6.7%, CEC 18 cmol kg^{-1}, base saturation 12.1%
C	10–25	Yellowish brown (10YR 5/6) moist color, sandy texture, massive structure, no reaction to NaF
2Ab1	25–80	Main color very dark gray (10YR 3/1) moist color, yellowish brown (10YR 5/6) moist secondary color, sandy loam texture, subangular blocky structure, strong reaction to NaF, pH 6.1, organic matter 5.1%, CEC 21 cmol kg^{-1}, base saturation 4.8%
2Ab2	80–110	Very dark grayish brown (10YR 3/2) moist color, loamy texture, subangular blocky structure, strong reaction to NaF

2.4.2 Central and Western Surfaces

These landscapes are formed by a close association of dissected surfaces located mainly at the central and western part of Great Tabular Cone. The representative profile corresponds to a soil classified as Hapludands (Soil Survey Staff 2006) (Fig. 2.13; Table 2.6). This soil has the following profile sequence: A/Bw/BC/C1/C2, a sandy loam texture in the first 50 cm; all horizons have strong reaction to NaF, horizon A has a very dark brown color and of granular to

Fig. 2.13 Profile of a soil classified as Typic Hapludands (left) located at Monte Olivo, Pedro Vicente Maldonado, Pichincha, in the central part of the Great Tabular Cone. These sites are used for silvopastoral systems to raise livestock for both milking and meet purposes and for agroforestry systems with crops like guava (*Psidium guajava* L.), cocoa (*Theobroma cacao*), naranjilla (*Solanum quitoense*), and coffee (*Coffea* sp.) (right) (IEE 2015)

Table 2.6 Profile characteristics of the soil classified as Typic Hapludands (IEE 2015)

Horizons	Depth (cm)	Description
A	0–20	Very dark brown (10YR 2/2) moist color, sandy loam texture, granular to subangular blocky structure, strong reaction to NaF, pH 5.7, organic matter 6.4%, CEC 16 cmol kg^{-1}, base saturation 21.4%, bulk density of 0.9 g cm^{-3}
Bw	20–50	Dark yellowish brown (10YR 4/4) moist color, sandy loam texture, subangular blocky structure, strong reaction to NaF, pH 6.1, organic matter 2.41%
BC	50–85	Dark yellowish brown (10YR 4/6) main moist color, yellowish brown (10YR 5/6) secondary color, massive to subangular blocky structure, strong reaction to NaF
C1	85–110	Yellowish brown (10YR 5/4) moist color, sandy loam texture, massive structure, strong reaction to NaF
C2	110–130+	Dark gray (10YR 4/1) main moist color, dark yellowish brown (10YR 4/4) secondary color. Sandy loamy texture, massive structure, strong reaction to NaF

subangular blocky structure. Underlying the horizon A is a Bw horizon of dark yellow brown color, angular blocky structure. The BC horizon is dark yellowish brown, sandy loam texture, and of massive to subangular blocky structure. The two C horizons have yellowish brown to dark gray colors with massive structure.

2.5 Old Deposition Plain

The old deposition plain is made of dissected and undulating reliefs that start at Quinindé at its northern part and move south as far as the line between Daule and Puebloviejo where it spreads into the recent alluvial plain. The plain

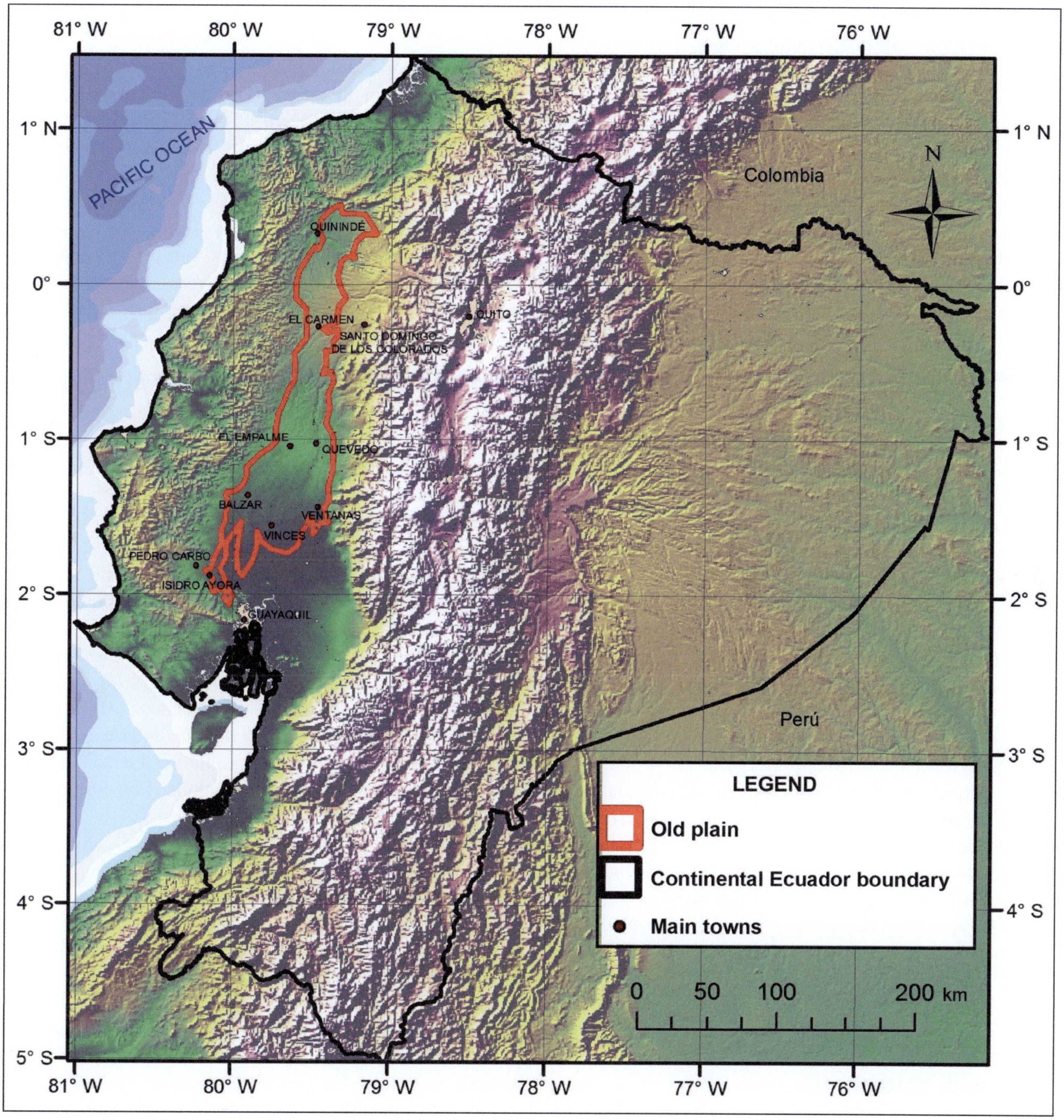

Fig. 2.14 Location of old alluvial plain with respect to continental Ecuador (adapted from Winckell and Zebrowski 1997)

borders to the east with the western Andes piedmont and to the west with the coastal sedimentary reliefs (Fig. 2.14). This zone is characterized for having dissected surfaces with wide or narrow round tops, slopes range from 2 to 40% with a maximum altitude of 15 m, but typically not exceeding 5 m (Winckell and Zebrowski 1997).

Genetically, this plain sits on old sediment deposits that were accumulated by the abundant contribution of the major rivers of the northern Highlands, the pyroclastic deposits of Andean volcanic material transported by the wind, and the subsequent boxing caused by the increased water runoff promoted by the melting of the Andean ice cap and the former hydrographic network that had more volume and strength that the current one due to one or more periods of glacial melting. The old plain, also known as dissected plain, has then been influenced by three sets of forming factors:

(a) dissection of deposited river sediments, (b) fluvial and pyroclastic surface shaping; and (c) climatic gradient, wet at the north and dry at the south. On account of the above-described conditions, the plain is divided into three homogeneous zones in terms of climate and pedogenesis: (a) surfaces fully covered by volcanic ash at the north; (b) zones with partial coverage of volcanic ash at the center; and (c) surfaces without pyroclastic ash cover to the south (León 2010; Winckell and Zebrowski 1997).

In climate terms, the old plain is located at very humid and humid megathermic tropical zone. The northern section and the strip that runs along the Andes piedmont receive an annual average rainfall ranging from 3,000 to 4,000 mm over a 195-day period with an annual deficit of <150 mm (<4 dry months), and with an average daily temperature ranging from 24 to 25 °C (CLIRSEN et al. 2009a). At the center (Vinces), the average annual rainfall reaches 1,500 mm and at the south (Isidro Ayora) only 970 mm products of 100 and 75 days of actual rain, respectively, which at the south represents an annual water deficit of 800 mm during the dry season (8–9 months) (Huttel 1997; Winckell and Zebrowski 1997). Recent survey data from Vinces ratifies that the average annual rainfall in that zone fluctuates from 1,400 to 1,650 mm and the average daily temperatures range from 25 to 27 °C (IEE 2015). A large segment of the central old plain has intermediate climate conditions characterized by an average annual rainfall of 2,000 mm, and an annual water deficit of 500 mm (5–7 dry months) (Winckell and Zebrowski 1997).

The old plain has the highest agricultural potential of the country which favors the production of major crops for domestic consumption and export. The crops that stand out are banana (*Musa* sp. div.), oil palm (*Elaeis guineensis*), cocoa (*Theobroma* cacao L.), mango (*Mangifera indica* L.), abaca (*Musa textiles* N.), plantain (*Musa paradisiaca*), corn (*Zea mays* L.), and rice (Oriza sativa) (Huttel 1997; INEC 2013; Pacheco 2009).

2.5.1 Northern Surfaces Fully Covered with Volcanic Ash

This zone corresponds to the northern old plain, which is the more humid section (>2500 mm annual rainfall), a condition that has promoted the formation of amorphous clays (allophane and imogolite) as a result of volcanic ash alteration and the marked loss of exchange bases (Winckell and Zebrowski 1997). Two general soils types are distinguished in this zone: (a) Slightly hydromorphic desaturated Andisols at the surface developed from recent ashes in an Udic and even Perudic moisture regime, but having slightly perhydrated fine-textured soils developed from old ashes deeper in the profile; and (b) soils fully covered by volcanic ash developed over red Paleosols in an Udic moisture regime.

Classification studies indicate that Andisols are the predominant soils in this zone (Fig. 2.15; Table 2.7). These soils have an effective depth >50 cm, loam, clayey loam, silty loam, or sandy loam surface texture, characteristic inherited from the ash coverage. Texture in deeper horizons continues in the loam range, except when they have underlaying Paleosols characterized by heavy clayey texture. Generally, soil drainage is good, and soil pH concentrates in the range from 6.0 to 7.0 in the upper horizons, very few profiles have a pH slightly higher than 7.0 (IEE 2015).

2.5.2 Zones with Partial Volcanic Ash Cover

This zone is located at the central part of the old plain, between El Empalme–Quevedo at the north and Balzar–Ventanas at the south. The ash coverage of this zone is thinner, relief is more pronounced, and there are outcrops of initially buried soils. The climate is drier compared to the areas fully covered by volcanic from the north, receiving <1,500 mm of annual rainfall. The soils are slightly more clayey than humid north and the content of amorphous clays decreases. The underlying buried soils are always saturated ferralitic soils (Winckell and Zebrowski 1997). The most important crops of this zone are banana (*Musa*sp div), plantain (*Musa paradisiaca*), cocoa (*Theobroma cacao*), and corn (*Zea mays* L.).The modal profile for this zone belongs to a soil classified as Typic Hapludolls (Soil Survey Staff 2006) (Fig. 2.16; Table 2.8).

2.5.3 Zones Without Ash Cover

This zone, which gradually diffuses into the low flood plain, has an Ustic moisture regime and, in consequence, is the driest of old plain. The soils developed exclusively from ancient alluvial sediments, since recent ash did not reach these areas because they are far away from the emission centers. Soils from the northeastern section located in more humid area, near the Andean western flank, are ferralitic; however, with the westward rainfall reduction as sites get away from Andean mountain range, soils become fersiallitic and then vertic (Winckell and Zebrowski 1997). Dominant soils in this region are Inceptisols and Alfisols (Table 2.9). The modal profile of this zone corresponds to a soil classified as Inceptic Haplustalfs (Soil Survey Staff 2006) (Fig. 2.17; Table 2.10).

Also in this zone, and generally throughout the old plain, there are valleys characterized for having alluvial deposits as substrate and flat slopes (0–2%) which tend to flood in the

Fig. 2.15 Profile of a soil classified as Typic Hapludands (left) at El Pital, Quevedo, Los Rios, with an A/Bw1/Bw2/C horizon sequence, located in a banana plantation on a dissected 3% slope surface (IEE 2015)

Table 2.7 Profile characteristics of the soil profile classified as Typic Hapludands (IEE 2015)

Horizons	Depth (cm)	Description
A	0–40	Very dark brown (10YR 2/2) moist color, loamy texture, granular structure, strong reaction to NaF, pH 6.6, organic matter 3.6%, CEC 23 cmol kg^{-1}, base saturation 43.9%, bulk density 0.68 g cm^{-3}
Bw1	40–60	Dark yellowish brown (10YR 4/6) moist color, loamy texture, subangular blocky structure, strong reaction to NaF, pH 6.8, organic matter 2.4%, CEC 22 cmol kg^{-1}, base saturation 33.1%
Bw2	60–105	Brown (10YR 5/3) moist color, loamy texture, subangular blocky structure, strong reaction to NaF, pH 6.4, organic matter 2.0%
C	105–110	Brown (7.5YR 4/3) moist color, clay loam texture, massive structure, strong reaction to NaF

rainy season, and then they are used to grow crops like upland rice and corn in dry periods (IEE 2012b). An example of the soils located in these valleys is the Typic Endoaquerts soil (Soil Survey Staff 2006) presented in Fig. 2.18 and Table 2.11 located at Mocache, Los Ríos, characterized by cracks, slickensides, high of 2:1 clay content (montmorillonite), aquic moisture regime and a water table located at 65 cm in the dry season, a condition that limits the effective depth. In addition, the soil has a clayey loam texture at the surface, a clayey texture in the lower horizons, and poor natural drainage (IEE 2015).

2.6 Recent Alluvial Plain

The recent alluvial plain lies between the Andean piedmont to the east, the old alluvial plain to the north, and the sedimentary reliefs of Manabí to the west, but it narrows

Fig. 2.16 Profile of a soil classified as Typic Hapludolls (left), located at Peñafiel, Mocache, Los Rios, on a 12% slope dissected area, on which cocoa has been planted (right) (IEE 2015)

Table 2.8 Profile characteristics of a soil profile classified as Typic Hapludolls (IEE 2015)

Horizons	Depth (cm)	Description
Ap	0–15	Very dark brown (10YR 2/2) moist color, silty loamy texture, grain structure, pH 6.9, organic matter 3%, CEC 33 cmol kg^{-1}, base saturation 93.4%
A	15–49	Very dark brown (10YR 2/2) moist color, silty loam texture, subangular blocky structure, pH 6.8, organic matter 0.6%
2AB	49–70	Very dark brown (10YR 2/2) main moist color, clayey loam texture, angular blocky structure, iron–manganese concretions, pH 6.3, organic matter 0.6%
2Bt1	70–93	Very dark brown (7.5YR 2.5/3) moist color, clayey texture, angular blocky structure, clay coatings, iron–manganese concretions, pH 6.6, organic matter 0.4%, CEC 14 cmol kg^{-1}, base saturation 93.1%
2Bt2	93–108+	Dark brown (7.5YR 3/4) moist color, clayey texture, massive to angular blocky structure, clay liners, pH 6.5, organic matter 0.3%

considerably between Guayaquil and the Masvale hills, and continues south forming only a narrow strip of about 10 km wide that extends as far as the border with Peru (Winckell and Zebrowski 1997) (Fig. 2.19). This is in one of the dry zones of the country; however, mean annual precipitation increases from west to east (1180–2215 mm), but it also decreases from north to south (1555–500 mm), presenting a dry period which lasts 8–9 months. Average annual temperatures are above 23 °C and annual temperature variations do not exceed 4 °C. This is a transition zone which tends to be a semi-humid environment when it approaches the Andean piedmont (Moreno 2001; Winckell and Zebrowski 1997).

These areas originated from quaternary detrital sediments of continental, alluvial, fluvial lacustrine, or fluvial-marine origin that filled a topographic depression created by the

Table 2.9 Percentage distribution of different soils at the subgroup level (Soil Survey Staff 2006), at the areas corresponding to Balzar, Colimes, Daule, Palestine, Santa Lucia, Puebloviejo, and Vinces, cantons located in the southern part of the old plain, characterized by the absence of a volcanic ash cover (IEE 2012b)

Subgroup	%
Vertic Haplustepts	42.95
Vertic Haplustalfs	15.91
Inceptic Haplustalfs	9.21
Typic Haplustalfs	8.19
Vertic Paleustalfs	6.71
Oxyaquic Vertic Haplustalfs	3.59
Typic Haplusterts	2.08
Typic Haplustepts	1.27
Dystric Haplustepts	1.22
Others	8.88

Fig. 2.17 Profile of a soil classified as Inceptic Haplustalfs located at Guayaca Puga, Puebloviejo, Los Rios, with an A/Bt/C horizon sequence (left). The site is located in a <12% slope dissected surface (right) (IEE 2015)

presence of a geological meridian trench. The landscape altitude is always <20 m a.s.l., but in most zones it only reaches 5 m, with a slight increase in altitude toward the piedmont. These are flat and monotonous areas, and the existent ondulations never are over 2 or 3 m in altitude; they are not easily perceptible (Winckell and Zebrowski 1997).

The formation of this alluvial plain started with the severe erosion processes in the Highlands that generated abundant

Table 2.10 Profile characteristics of the soil classified as Inceptic Haplustalfs (IEE 2015)

Horizons	Depth (cm)	Description
Ap	0–35	Very dark gray (10YR 3/1) moist color, with speckled reddish yellow (7.5YR 7/8) color, clayey loam texture, subangular blocky structure, pH 5.8, organic matter 4.1%, CEC 28 cmol kg^{-1}, base saturation 40.1%
Bt	35–53	Light yellowish brown (2.5Y 6/4) moist color, with speckled brownish yellow (10YR 6/8) color, clayey texture, massive structure, pH 6.7, organic matter 0.7%
C	53–125+	Light olive brown (2.5Y 5/4) main moist color, dark grayish brown (2.5Y 4/2) secondary color, with speckled reddish yellow (7.5YR 7/8) color, clayey texture, massive structure

Fig. 2.18 Profile of a soil classified as Typic Endoaquerts (left) located at Garza Grande, Mocache, Los Ríos, with an Apss/ACssg horizon sequence. The photograph was taken in the dry season (IEE 2015)

Table 2.11 Profile characteristics of the soil profile classified as Typic Endoaquerts (IEE 2015)

Horizons	Depth (cm)	Description
Apss	0–15	Dark gray (5Y 4/1) moist color, with speckled red (2.5YR 4/6) color, clayey loam texture, blocky structure, pH 5.5, organic matter 5.2%, CEC 29 cmol kg^{-1}, base saturation 69.5%
ACssg	15–55	Gray (5Y 5/1) moist color, clayey texture, massive to angular blocky structure, pH 6.4, organic matter 2.7%

sediments which were washed by rivers that finally reached the old alluvial plain or the dissected cones. These rivers spilled over this great plain generating clay and silt particle settling. The sediment accumulation affects the Gulf of Guayaquil where the fluvio-marine environment of the Guayas River estuary loses ground because the sediment contributions are higher than the spaces generated by the continuous sinking of actual subsidence structure.

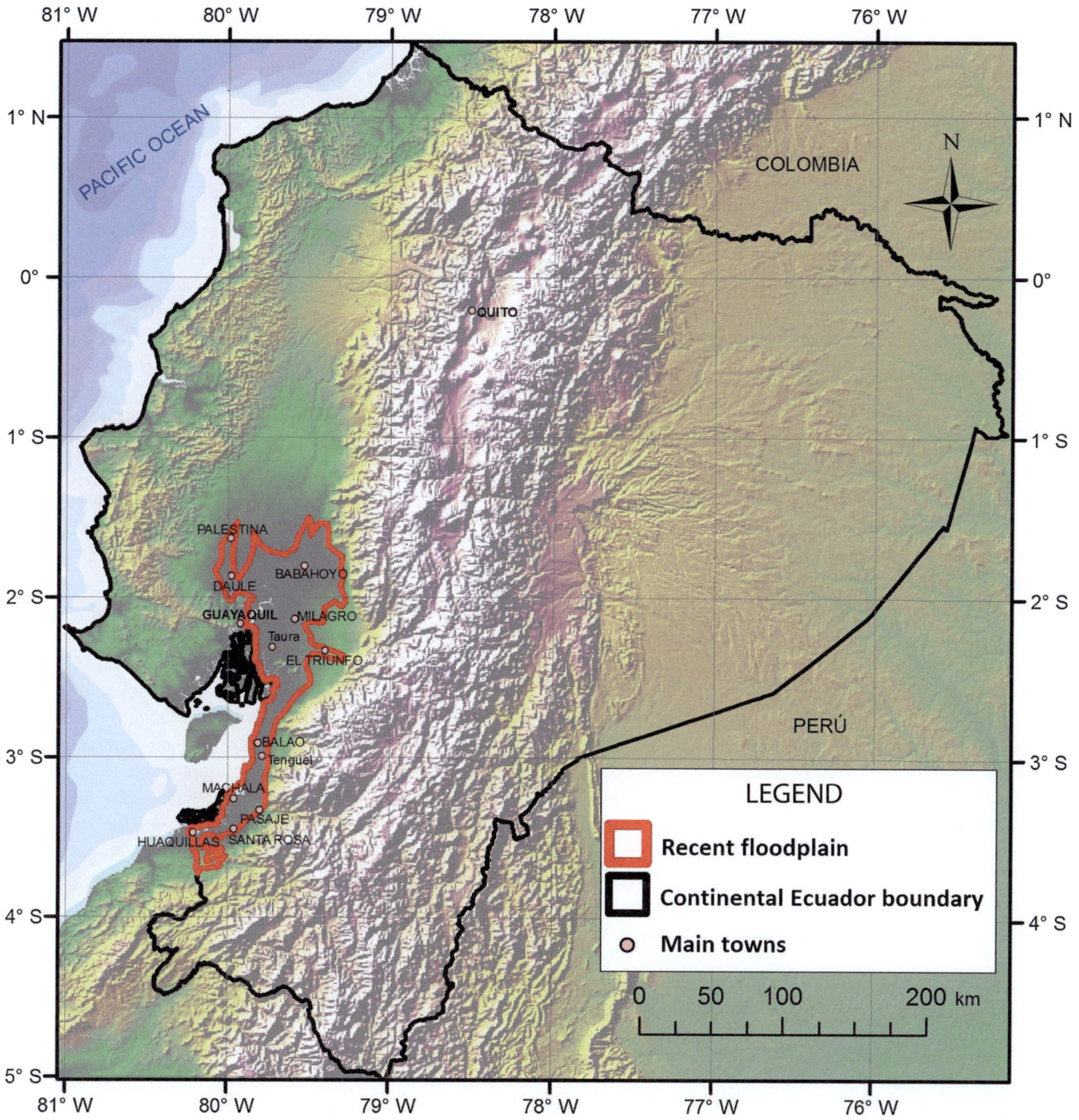

Fig. 2.19 Location of the recent alluvial plain with respect to the continental Ecuador (adapted from Winckell and Zebrowsky 1997)

Consequently, the soils of the recent alluvial plain are remodeled by periodic floods that come with the rainy seasons (Tapia 2012; Winckell and Zebrowski 1997). It is important to note that the lower Guayas River basin is immersed within this plain, which is the most important agricultural zone of the country. This basin is formed by the lower sub-basins of the Daule, Vinces, Babahoyo, Chimbo, Bulubulu, Taura, and Churute rivers (Lasso et al. 2010; Sevillano 2010).

As a result of recent formation factors, the low alluvial plain is subject to flooding, and in general, soils are characterized by a homogeneous clayey texture to an appreciable depth. Those soils with shallow water table show horizons of very sharp colors marked by hydromorphism. Areas near the fluvio-marine environment can have saline and totally flooded soils. The heavy clayey soils (>60% clay) located at the bottom of plain are black when they are not submerged part of the year, but soils at lower sites are flooded much of the year and present olive gray color, usually, slightly acidic pH (6.0–7.0), and a base saturation >50% (Moreno 2001; Winckell and Zebrowski 1997).

The low alluvial plain is characterized by the intensive agricultural use. The floodplain at the lower levels has important areas for patty rice production inundated by natural flooding in the rainy season and by pumped irrigation in the dry season. The highest areas, with less flooding risk, are used for sugarcane or banana plantations (IEE 2012b; Winckell and Zebrowski 1997).

2.6.1 Flat Plain

The flat plain is characterized by gradients that do not exceed 2%, and with no relative incline, reason why these are areas frequently flooded. Predominant soils are Vertisols (Soil Survey Staff 2006) (Table 2.12) which are characterized by heavy texture with the presence of expansive-type clays that form cracks in the dry season and expand in the rainy season. Due to the low relative inclination, soils are poorly drained and show a massive structure (CLIRSEN et al. 2009a).

The main limitation of this type of soils is the poor physical characteristics that generate conditions of poor drainage, low water and air conductivity, heavy textures, poor structural development and porosity, water saturation, and high water tables (Table 2.13). These are very difficult tillage soils with severe limitations to grow crops, except patty rice (Fig. 2.20). The cost of corrective drainage to habilitate these soils for other crops is very high (Mejía 1997).

The stratification and insertion processes of the fluvial and marine deposits promote salinity, condition which is evident in the profile presented in Fig. 2.21 and Table 2.14, showing changes in electrical conductivity ranging from 0.17 to 11.0 dS m^{-1} below 55 cm in the profile.

2.6.2 Banks and Alluvial Dams

The banks and alluvial dams area is located primarily at the Daule and Babahoyo river banks, but also along the sinuous border of the hydrographic system of the flat plain (Fig. 2.22). These are depositional areas, with a 0–5% dominant slope, and a relative gradient that varies from 0 to 5 m (CLIRSEN et al. 2009d). Banks and dikes play the important environmental role of protecting riverbanks with a tropical tree vegetation cover like mango which provides shade to coffee, and cocoa. Banana, citrus, and other subsistence crops are also grown in these areas. Banks and alluvial dams also protect adjacent land against river flooding (Winckell and Zebrowski 1997). Soils classified as Mollisols are predominant in the banks and alluvial dams (Figs. 2.22 and 2.23), followed by Entisols and Inceptisols (Table 2.15).

A good portion of the alluvial banks has been altered to meet the requirements for patty rice and sugarcane production. At the agricultural area of Milagro, between 1983 and 2010, soils classified as Fluventic Hapludolls, developed over banks and alluvial dams, were altered and degraded to Fluventic Eutrudepts, mainly by human intervention in response to the price decline of crops such as cacao and coffee, and the increment of patty rice and sugarcane. Farmers literally eliminated the mollic epipedon in order to find clayey layers that allow production of crops such as patty rice. It is estimated that the disturbed soils cover an area of 5,970 ha (Montúfar et al. 2010).

These soils are characterized by having, in general, a loamy texture within the first 50 cm, soils are friable and have good drainage, and therefore easy to manage. Similarly, chemical properties are favorable for agricultural

Table 2.12 Percentage distribution of different soil types, at the subgroup level (Soil Taxonomy 2006) at the flat plain (IEE 2012a)

Subgroup	%
Typic Hapluderts	58.26
Entic Haplusterts	10.70
Aquic Hapluderts	10.67
Fluventic Hapludolls	5.76
Typic Haplusterts	5.17
Typic Udifluvents	5.05
Aquic Haplustepts	3.25
Aeric Endoaquepts	0.67
Udic Haplusterts	0.35
Vertic Eutrudepts	0.07

Table 2.13 Profile characteristics of a soil classified as Typic Hapluderts (IEE 2015)

Horizons	Depth (cm)	Description
Ap	0–30	Dark gray (10YR 4/1) and black (10YR 2/1) moist color, clayey texture, wedge-shaped mixed with massive structure, pH 6.7, organic matter 1.5%, CEC 27.6 cmol kg^{-1}, base saturation 98.7%
Css1	30–80	Very dark gray (5Y 3/1) moist color, clayey texture, wedge-shaped structure mixed with massive structure, many friction faces, pH 7.2, organic matter 0.7%, CEC 31.8 cmol kg^{-1}, base saturation 97.9%
Css2	80–130	Main color Dark gray (5Y 4/1) main moist color and olive gray (5Y 5/2) secondary moist color, clayey texture, massive structure, pH 7.3, organic matter 0.4%, CEC 42.3 cmol kg^{-1}, base saturation 98.1%

Fig. 2.20 Profile of a soil classified as Typic Hapluderts (left) located at Balsa, Colimes, Guayas, in a flat area where rice was grown (right). All horizons have >40% clay (IEE 2015)

production, pH is not far from neutrality, organic content is >2%, and base saturation is >50% (CLIRSEN et al. 2009d). The modal profile for this type of soils is presented in Fig. 2.23 and Table 2.16.

2.6.3 Slightly Wavy Areas with the Presence of Water

This zone is located in the area of influence of the Babahoyo and Daule rivers, water bodies that deposit sediments generating an association of very smooth ondulations (maximum 3–5 m), and permanently flooded depressions slightly perceptible to the eye. This is the typical manifestation of an undulating molding which only shows the peaks emerging from the water. Sites within this zone are characterized by 2–5% slopes if they are natural and 0–2% if they have been mechanized. Vegetation cover consists of patty rice, pastures, and natural vegetation (Fig. 2.24) (CLIRSEN et al. 2009c; Moreno 2001; Winckell and Zebrowski 1997). Semidetailed soil surveys conducted in this zone demonstrated that the dominant soils are Vertisols and Inceptisols, but they also reported the presence of Entisols and Mollisols in lesser extent (Table 2.17) (IEE 2012a).

In general, soils from this landscape are clayey and shallow. These soils are represented in associations due to the difficulty to separate them cartographically, for example, soils that remain flooded most of the year (depressions) have gley colors (olive green) and they have been classified as Vertic Endoaquepts, because they have an aquic moisture regime; in contrast, soils at the emerged tops have black colors and they have been classified as Udic Haplusterts

Fig. 2.21 Profile of a soil profile classified as Aeric Endoaquents (left) located at La Esperanza, Montalvo, Los Rios, in a flat area where patty rice was grown (right). The water table at this site is at a depth of 70 cm, and a layer of dark greenish gray color shows at 10 cm (IEE 2015)

Table 2.14 Profile characteristics of a soil classified as Aeric Endoaquents (IEE 2015)

Horizons	Depth (cm)	Description
Ap	0–10	Olive brown (2.5Y 4/3), and yellowish red mottled (5YR 4/6) moist colors, silty clay texture, subangular blocky structure, pH 6.6, organic matter 2%, CEC 18 cmol kg^{-1}, base saturation 69.9%, electrical conductivity 0.28 dS m^{-1}
C1	10–30	Dark greenish gray (GLEY 1 5GY 4/1) moist color, clayey texture, crumbly structure, pH 7.1, organic matter 1.9%, CEC 19 cmol kg^{-1}, base saturation 73%, electrical conductivity 0.2 dS m^{-1}
2C2	30–40	Yellowish brown (10YR 5/4), and dark greenish gray (GLEY 1 5GY 4/1) mottled moist colors, loamy texture, pH 7.4, CEC 14 cmol kg^{-1}, base saturation 78.9%, electrical conductivity 0.16 dS m^{-1}
2C3	40–55	Yellowish brown (10YR 5/4), dark greenish gray mottled (GLEY 1 5GY 4/1) moist colors, sandy loam texture, crumbly structure, pH 6.8, organic matter 0.05%, electrical conductivity 0.17 dS m^{-1}
3C4	55–70	Light olive brown (2.5Y 5/3), and strong brown mottled (7.5YR 5/8) moist colors, loamy texture, pH 7, organic matter 0.4%, electrical conductivity 11.0 dS m^{-1}

(Fig. 2.25). There are also poorly drained, heavy textured soils that remain saturated part of the year (usually from January to May) (Fig. 2.26; Table 2.18). These soils tend to be slightly acid (pH 6.0–6.5), slight saline (electrical conductivity >2.0 and <4.0 dS m^{-1}), and have a base saturation >50% (CLIRSEN et al. 2009c; Moreno 2001; Winckell and Zebrowski 1997).

2.7 Piedmont of the Western Andean Mountain Range

This unit is located between the western Andean flanks to the east and the alluvial plain to the west (Fig. 2.27). Its position immediately at the foot of the Andean flanks originates the name. The unit was formed by the sweeping of

Fig. 2.22 Soil vegetation relationship in the recent alluvial plain. At the center, a 2009 KOMPSAT image (4,2,1—RGB band combination). On the left, a profile of a soil classified as Fluventic Hapludolls located at El Hacha, Vinces, Los Ríos developed over either a bank or an alluvial dam (**a** in the image) used for perennials such as cocoa and tropical fruits or crops like corn. On the right, a profile of a soil classified as Vertic Haplustepts located at Junquillo, Vinces, Los Ríos in an area slightly dissected area (**b** in the image) covered with cultivated pastures (Sevillano 2010)

Table 2.15 Percentage distribution of different soil types, at a subgroup level (Soil Taxonomy 2006), from the banks and alluvial dams (IEE 2012a)

Subgroup	%
Fluventic Hapludolls	65.20
Fluventic Eutrudepts	15.30
Mollic Udifluvents	12.90
Typic Hapludolls	3.02
Typic Udifluvents	1.76
Vertic Eutrudepts	0.59
Typic Ustipsamments	0.43
Udic Ustifluvents	0.37
Other	0.43

colluvial–alluvial deposits provoked by the melting of Quaternary glaciers which moved at high-speed drawing conical figures on the surface of the plain; thus, minerals are varied both in composition and in size and shape. The diagram presented in Fig. 2.28 describes the direction and force of sediment deposits (MIDENA et al. 2013b; Winckell and Zebrowski 1997).

There is a marked difference between the Piedmont northern landscapes and those located at the center and south. The northern area is completely covered with

Fig. 2.23 Profile of a soil profile classified as Fluventic Hapludolls (left) located at Recinto San Ramón, Vinces, Los Rios, developed over several layers of loose friable deposits in an alluvial bank with an A/Bw/C1/C2/2Ab/2C1/2C2 horizon sequence. The site is cultivated with cocoa (right) (IEE 2015)

Table 2.16 Profile characteristics of a soil classified as Fluventic Hapludolls (IEE 2015)

Horizons	Depth (cm)	Description
A	0–30	Dark brown (10YR 3/3) moist color, loamy texture, granular structure, pH 6.4, organic matter 2.9%, CEC 17.3 cmol kg^{-1}, base saturation 98.7%
Bw	30–58	Light olive brown (2.5Y 5/6) moist color, loamy texture, granular structure, pH 7.0, organic matter 1.9%, CEC 19.9 cmol kg^{-1}, base saturation 98.4%
C1	58–76	Light olive brown (2.5Y 5/4) dry color brown mottled (7.5YR 4/4) when moist, silty clay loam texture, pH 7.3, organic matter 1.8%
C2	76–90	Light olive brown (2.5Y 5/4) dry color with dark brown mottled (7.5YR 3/4) moist color, silty clay texture, pH 7.2, organic matter 1.6%
2Ab	90–97	Main color Dark grayish brown (10YR 4/2) main moist color, light olive brown (2.5Y 5/3) secondary color when moist, dark brown (7.5YR 3/4) mottled dry color, clayey texture, granular structure, pH 7.3, organic matter 1.5%
2C1	97–120	Light olive brown (2.5Y 5/4) moist color, dark yellowish brown (10YR 4/6) mottled moist color, clayey texture
2C2	120–153	Light olive brown (2.5Y 5/4) dry and moist color, yellowish brown mottled (10YR 5/8) dry and moist color, clayey texture

pyroclastic material, condition that smooths out the geoforms and provides particular physical and chemical characteristics to the soils. The altitude decreases as the unit moves southward, from around 1000 m o.s.l. in Alluriquín at the north to 20 m o.s.l. where the Piedmont enters in contact with the alluvial plain at the south. Similarly, there is a reduction in the amount of average annual precipitation that varies from 2660 mm at the north to 500 mm at the south. In addition, there is also a secondary precipitation gradient from east to west due to orographic blocking of air circulation; the highest average annual precipitation occurs close to the Andean flanks (2290 mm at Bucay) and the lowest at the western edge of the Piedmont (1440 mm at San Carlos) (Winckell and Zebrowski 1997). There are two clearly defined subunits within the Piedmont: (a) Dejection cones and (b) scattered cones.

2.7.1 Dejected Cones

The dejected or alluvial cones are located in direct contact with the Andean flank (Fig. 2.28) on slopes ranging from 12 to 25% characterized by dissected surfaces due to the long time under the action of remodeling processes. This zone was formed by glacial melting which broth detrital, alluvial, and torrential deposit of pebbles, blocks, and gravel in a sandy-clayey matrix that extended radially downslope as a result of the sharp gradient decline from the point where the

Fig. 2.24 Overview of a slightly ondulating area with vertic soils located at the Cooperativa Vista Alegre, Samborondón, Guayas. At the bottom, the flooded soil covered with hydrophilic plants (*Eichhornia* sp.), and on the top the non-flooded soil with trees (CLIRSEN et al. 2009c)

Table 2.17 Percentage distribution of different soil types, at subgroup level in the slightly ondulating area (IEE 2012a)

Subgroup	%
Typic Haplusterts	24.92
Typic Hapluderts	19.40
Vertic Eutrudepts	9.12
Vertic Endoaquepts	9.11
Udic Haplusterts	5.57
Others	31.88

flow of water leaves the rugged area (MIDENA et al. 2013b). Dominant soils in this area are Andisols and Alfisoles (Table 2.19).

The soils in the northern part of this unit are covered by a thick layer of volcanic ash which is 6 m deep at the northern part (areas close to the emission centers) and can be only 10 cm deep at the northern central part. The predominant soils of this area are Andisols characterized by low bulk density (<0.9 g cm^{-3}) and high moisture retention (Table 2.20). Rainfall differences within the dejected cone are responsible for the unit particular soil characteristics. At the more humid zone, closer to the ash emission centers, desaturated Andisols (<50% base saturation) have been developed. At the same humid zone, but far away from the emission centers, perhydrated Andisols (water retention >100%) were developed, and at the drier zones saturated Andisols (>50% base saturation) are evident (IEE 2015; Winckell and Zebrowski 1997; Zebrowski and Sourdat 1997).

The soils in the central and southern parts of the dejected cones are characterized by having clayey-stony texture, because they were originally developed from stony deposits. These soils have been leached by high rainfall, conditions which have promoted the development of ferralitic soils at the more humid zones and fersiallitic at the drier zones. This area is dominated by soils classified as Alfisols (Table 2.20).

An example of the soils from the northern area is the soil classified as Typic Hapludands (Soil Survey Staff 2006) (Fig. 2.29; Table 2.21) characterized by having moderate depth (<100 cm), sandy loam texture, slightly acid (pH 6.2), high organic matter content (4.3%), and low base saturation (31.1%) (MIDENA et al. 2013b).

An example of a soil from the central and southern part of the dejected cones is presented in Fig. 2.30 and Table 2.22. This soil, classified as Mollic Paleudalfs (Soil Survey Staff 2006), has a shallow effective depth (<50 cm) due to the heavy texture (>50% clay) of the deeper horizons which prevents root penetration. Weighted-average pH, organic matter content, and base saturation in the first 47 cm are 6.7, 2.2, and 54.7%, respectively, conditions of a soil with high natural fertility (IEE 2015).

2.7.2 Dispersion Cones

This zone is located immediately below the dejected cones. The alluvial deposits were dispersed over larger areas and modeled a lower gradient relief with mostly flat surfaces of

Fig. 2.25 Profile of a soil classified as Udic Haplusterts (left) located at El Hormiguero, Samborondón, Guayas, where the characteristic blocks of a Vertisol are evident. Patty rice is the main crop grown in these soils (right) (IEE 2015)

Fig. 2.26 Profile of a soil profile classified as Aquertic Hapludolls (left) located at El Mango Morado, Babahoyo, Los Rios, cultivated with patty rice (right) (IEE 2015)

Table 2.18 Profile characteristics of a soil classified as Aquertic Hapludolls (IEE 2015)

Horizons	Depth (cm)	Description
Ap	0–18	Black (10YR 2/1) moist color with strong presence brown mottling (7.5YR 5/8), clayey texture, granular structure, pH 5.5, organic matter content 3.2%, CEC 21.8 cmol kg^{-1}, of base saturation 97.2%, of electric conductivity 1.5 dS m^{-1}
Ag	18–40	Dark greenish gray (Gley 1 4/5GY) moist color with red mottling (2.5YR 4/8), olive yellow (2.5Y 6/8) secondary moist color, silty loam texture, pH 6.9, organic matter 0.7%, CEC 32.8 cmol kg^{-1}, base saturation 98.8%, electric conductivity 4.84 dS m^{-1}
Bg	40–90	Olive gray (5Y 4/2) moist color with light olive brown mottling (2.5Y 5/6), clayey texture, pH 7.7, organic matter 1.2%, CEC 41.1 cmol kg^{-1}, base saturation 98.8%, electric conductivity 3.2 dS m^{-1}
Cg	90–100+	Gray (2.5Y 4/1) moist color, silty loam texture, massive structure

large undulating slopes that do not exceed 12%. Deposits are sandy-rocky at the top and sandy-silty at lower part as the relief gradually reaches the recent alluvial plain.

At the north side of this unit, covered with volcanic ash, the humid climate has promoted the development of desaturated Andisols; however, in the less humid climate, soils are saturated Andisols. The southern zone of this unit, drier and without volcanic ash cover, presents sandy-silty soils with gravel at the higher part, and silty clay soils at the lower part which gradually merges with the recent alluvial plain. Deposits in this unit have not completely aged due to the deposits younger age and the drier environment; this condition has promoted the development of 2:1 clays (Winckell and Zebrowski 1997).

A typical soil of the northern part of the dispersed cones is presented in Fig. 2.31 and Table 2.23. This soil, classified as Typic Hapludands (Soil Survey Staff 2006), has good physical characteristics (loamy within the profile first 40 cm) which can be tilled for agricultural use. Weighted averages of the chemical characteristics for the first 40 cm are 5.3, 4.1, and 23.8% for pH, organic matter content, and base saturation, respectively (MIDENA et al. 2013b).

The soil profile presented in Fig. 2.32 and Table 2.24 represents the southern part of the dispersed cones with no volcanic ash cover. This soil, located near Naranjito, classified as Mollic Ustifluvents (Soil Survey Staff 2006) has >100 cm effective depth; texture of the first 40 cm of the profile varies from loam to clayey loam, but texture is sandy loam from 40 to 100 cm. Because soils are in an Ustic moisture regime, base saturation is >50% (CLIRSEN et al. 2009b; Zebrowski and Sourdat 1997).

2.8 Sedimentary and Fluvio-Marine Reliefs

The sedimentary and fluvio-marines reliefs are located in the central littoral regions of the Manabí and Santa Elena Provinces. The development of these reliefs is consequence of the direct action of the sea and the various marine processes related with the transgression–regression marine events, and the old and new deposits, which formed the littoral plains, beach ridges, and different levels of marine plateaus. These landscapes include medium and low seasides, with minor cliffs and large front straight beaches in sectors near Manta and the entire periphery of the Santa Elena Peninsula where different quaternary transgressions were deposited (Fig. 2.33) (Collot et al. 2009; Winckell and Zebrowski 1997). The presence of herbaceous and shrub vegetation is typical of the desert thorny scrub of the dry tropical areas. The average annual temperature is >23 °C (Winckell and Zebrowski 1997; Zebrowski and Sourdat 1997).

2.8.1 Sandstone Calcareous Plateaus

The calcareous plateaus correspond to areas of dry tropical to subdesert climate that receives directly the drying effects of the Humboldt cold current; the average annual precipitation varies from 240 mm near the coast to 400 mm inland, reaching 800 mm on the plateau summits. This zone covers mainly the uneven coastal perched, rolling, and dissected plateaus located over recent sediments. These calcareous sandstone plateaus extend in an inclined plane from 80–100 to 320–360 m o.s.l. from the coastal line as far as Montecristi inland. The continuity of all of these surfaces is due to the existence of recent marine transgressions that deposited a succession of detrital sediments, probably of Pleistocene age, that were affected by Quaternary tectonic activity. They are located exactly where the Carnegie submarine ridge collides with the continent, condition which allowed that deposits of these transgressions arise above 350 m o.s.l. Plateau composition is defined by alternating layers of moderately cemented sands and sandstones and fossiliferous calcareous sandstone having in the middle a well-cemented basal conglomerate, corresponding to the Tablazo formation. On the slopes of ravines, formations conformed by rocky

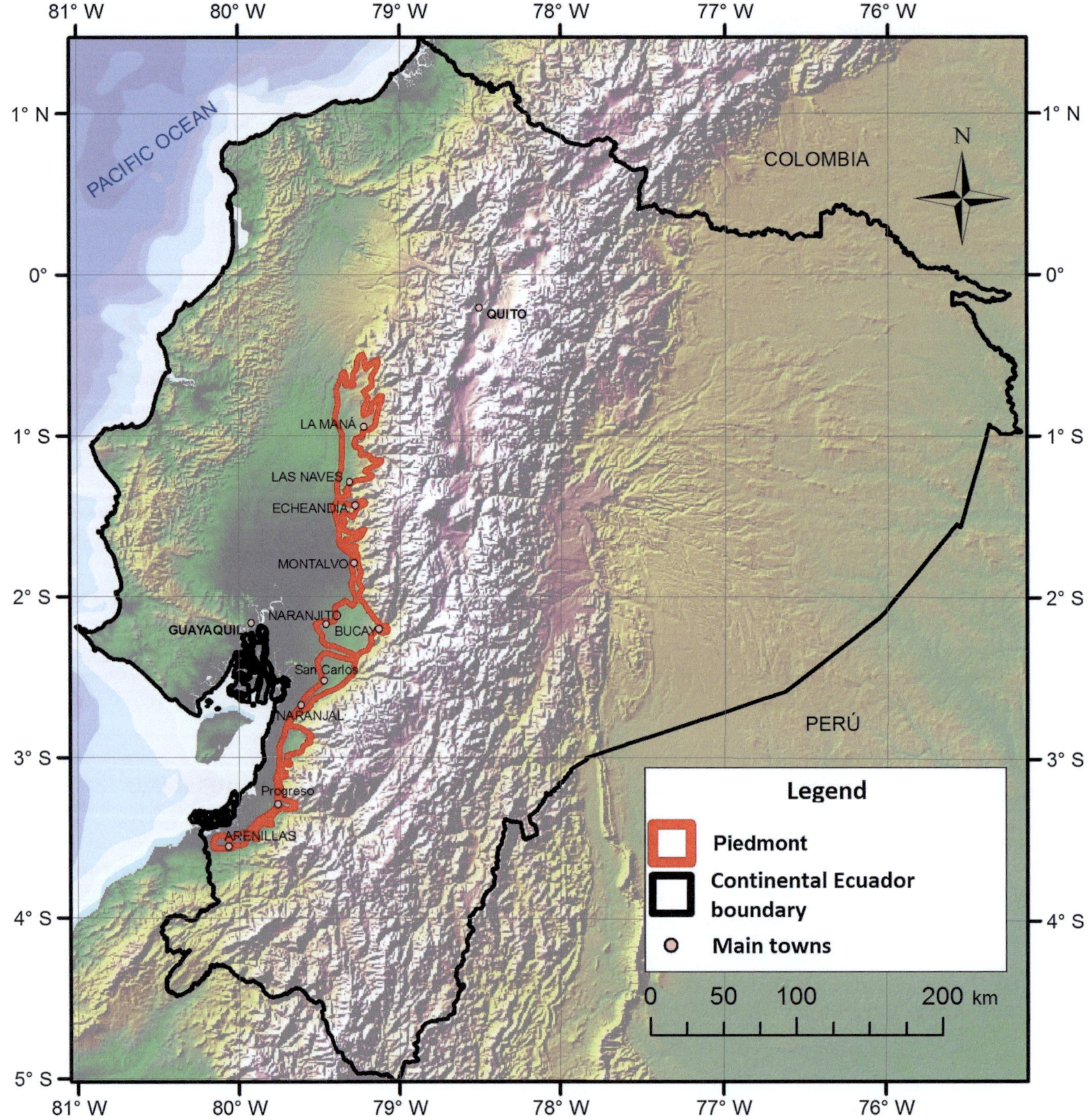

Fig. 2.27 Location of the western Andean Piedmont with respect to continental Ecuador (adapted from Winckell and Zebrowski 1997)

outcrops are alternated over more consolidated banks of fine to medium grain compacted calcareous sandstones, lumaquels and, in many cases, recent sand deposits (Collot et al. 2009; Winckell and Zebrowski 1997; Zebrowski and Sourdat 1997).

The soils in this zone are clayey, vertic, moderately profound, with a textural discontinuity between 10 and 50 cm (top horizon poor in clay and the subjacent more clayey) which often have accumulations of calcium carbonate and even gypsum, due to pedoclimate changing from dry to very dry (Winckell and Zebrowski 1997; Zebrowski and Sourdat 1997).

The modal profile of this zone corresponds to a soil classified as a Typic Haplustalfs (Soil Survey Staff 2006) (Fig. 2.34; Table 2.25). Morphologically, this soil presents the following horizon sequence: A/AE/Bt/C with evident

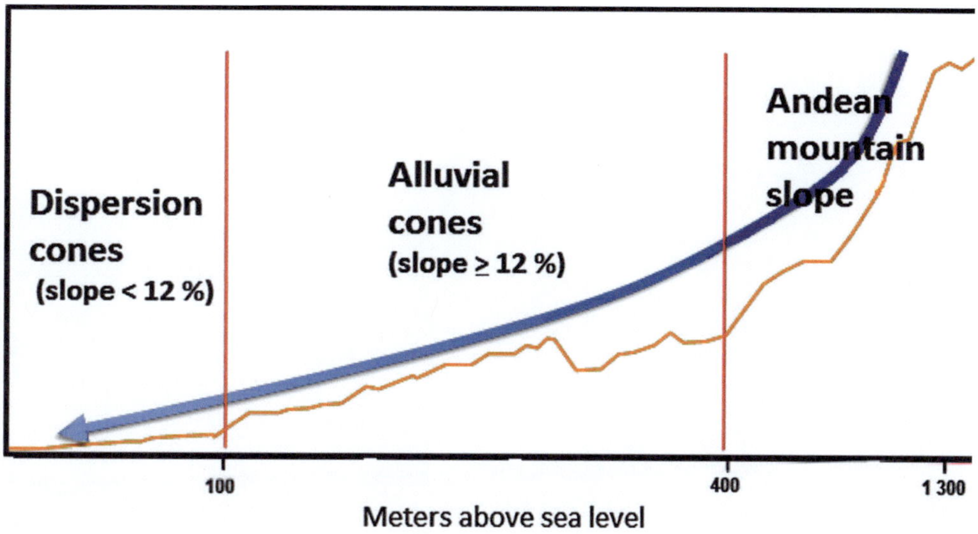

Fig. 2.28 Surface contour of Western Andean Piedmont. The brown line represents the ground surface and the blue arrow represents the direction and strength of the deposits, a darker line indicates a greater force (MIDENA et al. 2013b)

Table 2.19 Percentage distribution of the different soil types at the order level in the dejected cones (IEE 2012a)

Order	%
Andisols	44.04
Alfisols	45.28
Inceptisols	7.56
Entisols	2.97
Mollisols	1.15

Table 2.20 Percentage distribution of the different soil types, at subgroup level, at the northern and central southern sectors of the dejection cones (IEE 2012a)

Subgroup	%
Northern sector	
Typic Hapludands	89.73
Thaptic Hapludands	10.27
Central and southern sectors	
Typic Hapludalfs	42.97
Typic Hapludands	18.42
Inceptic Hapludalfs	13.49
Humic Dystrudepts	5.39
Typic Udorthents	4.32
Ultic Hapludalfs	3.67
Typic Dystrudepts	3.39
Mollic Paleudalfs	3.35
Andic Paleudalfs	2.53
Humic Eutrudepts	1.53

albic and argillic horizons. This type of soil is found at the <5% slopes landforms of the marine plateau surfaces characterized by an Ustic moisture regime and an isohyperthermic temperature regime (CLIRSEN et al. 2011a).

2.8.2 Littoral Strip

This zone is located at the western part of the Santa Elena peninsula, as plains and undulations characterized by nearly

Fig. 2.29 Profile of a soil classified as Typic Hapludands (left) located at Manguila, Valencia, Los Ríos, on a 25–40% slope. These soils are suitable for tree plantations such as rubber (right) (IEE 2015)

Table 2.21 Profile characteristics of a soil classified as Typic Hapludands (IEE 2015)

Horizons	Depth (cm)	Description
Oe	0–5	Black (10YR 2/1) moist color, loamy sand texture, granular structure
A	5–15	Main color Very dark gray (10YR 3/1) main moist color, very dark gray brown (10YR 3/2) secondary moist color, sandy loam texture, granular to subangular blocky structure, slight reaction to NaF, pH 6.2, organic matter 4.3%, CEC 20 cmol kg^{-1}, base saturation 31.1%
Bw	15–35	Very dark grayish brown (10YR 3/2) main moist color, dark brown (10YR 3/3) secondary moist color, loamy texture, subangular blocky structure, strong reaction to NaF
C1	35–60	Dark brown (10YR 3/3) moist color, sandy clay loam texture, massive structure, strong reaction to NaF
C2	60–90	Dark brown (10YR 3/3) main moist color, dark grayish brown (10YR 4/2) secondary moist color, sandy clay loam texture, massive structure, strong reaction to NaF

flat reliefs of clayey sedimentary to sandy hills. This peninsula is fully exposed to the drying effects of the Humboldt cold current to the point that the tip of Salinas could be considered the final part of the Peruvian coastal desert; therefore, rainfall is very scarce. Average annual rainfall is <100 mm at Salinas and up to 400 mm at Playas. This littoral landscape of desert like climates is divided into three sub-landscapes: (a) Eroded hills of sedimentary base, (b) recent low undulations, and (c) littoral deposits.

Eroded Hills from Sedimentary Base

This type of landscape is located in two well-defined distinct sectors of the Ecuadorian littoral, behind the Santa Elena coast to the north, and at either side of Ancon to the south. This zone is characterized by reliefs <50 m o.s.l., with flat plateaus truncated by Quaternary marine abrasion, covered or not by marine deposits. In general, this is an area with steep slopes (>40%), with variable dissection where the poor

Fig. 2.30 Profile of a soil profile classified as Mollic Paleudalfs (left) located at hacienda La Independencia, Ventanas, Los Rios, with an A/Bt/BC/C horizon sequence planted with cocoa (right) (IEE 2015)

Table 2.22 Profile characteristics of a soil classified as Mollic Paleudalfs (IEE 2015)

Horizons	Depth (cm)	Description
Ap	0–25	Very dark grayish brown (10YR 3/2) dry color, clayey texture, granular to subangular blocky structure, pH 6.7, organic matter 3.1%, CEC 24 cmol kg^{-1}, base saturation 60.7%
Bt	25–47	Very dark grayish brown (10YR 3/2) dry color, clayey texture, massive to subangular blocky structure with clay liners, pH 6.8, organic matter 1.3%, CEC 28 cmol kg^{-1}, of base saturation 48.0%
BC	47–62	Dark brown (7.5YR 3/3) moist color, clayey texture, massive to subangular blocky structure, pH 6.5, organic matter 0.5%
C	62–80+	Dark brown (7.5YR 3/4) moist color, clayey texture, massive structure, presence of fine gravel

xerophytic vegetation does not cover and stabilizes the soil, which is easily eroded by the infrequent rainfall events forming gullies (Winckell and Zebrowski 1997; Zebrowski and Sourdat 1997).

A characteristic soil of this zone, classified as Calcic Petrocalcids (Soil Survey Staff 2006), is presented in Fig. 2.35 and Table 2.26. These incipiently developed soils present excess of carbonates, gypsum, or salts that drastically limit crop production. The main feature of these soils is the presence of a calcic horizon over a petrocalcic horizon where carbonates have accumulated forming a hard horizon. The reference profile (Fig. 2.35) is found on steep slopes (40–70%), has sandy-clayey loam texture on the surface, and clayey loam in depth, and has the following horizon

Fig. 2.31 Profile of a soil classified as Typic Hapludands (left), located at La Experiencia de Chipe, Valencia, Los Ríos. The soil horizon sequence is Ap/Bw/C/2Ab/2C, showing evidence of two events of volcanic ash deposition. This <12% slope relief is cultivated with soybeans (right) (IEE 2015)

Table 2.23 Profile characteristics of a soil classified as Typic Hapludands (IEE 2015)

Horizons	Depth (cm)	Description
Ap	0–20	Main color Very dark gray (10YR 3/2) main moist color, yellowish brown mottling (10YR 5/8) moist color, loamy texture, granular to subangular blocky structure, strong reaction to NaF, pH 5.8, organic matter 5.3%, CEC 21 cmol kg^{-1}, base saturation 25.4%
Bw	20–40	Very dark grayish brown (10YR 3/2) moist color, sandy loam texture, subangular blocky structure, strong reaction to NaF, pH 6.5, organic matter 3.81%, CEC 19 cmol kg^{-1}, base saturation 29.2%
C	40–65	Dark yellowish brown (10YR 3/4) moist color, silty loam texture, porous massive structure, strong reaction to NaF
2Ab	65–90	Very dark brown (10YR 2/2) moist color, silty clay loam texture, massive to subangular blocky structure, strong reaction to NaF
2C	90–100	Dark yellowish brown (10YR 3/6) moist color, silty clay loam texture, massive structure, strong reaction to NaF

Fig. 2.32 Profile of a soil classified as Mollic Ustifluvents (left), located at the San Carlos sugar plantation, Naranjito, Guayas, with the following horizon sequence: Ap/A/AC/CA/C. This zone holds the largest sugarcane area of the country (right) (IEE 2015)

Table 2.24 Profile characteristics of a soil classified as Mollic Ustifluvents (IEE 2015)

Horizons	Depth(cm)	Description
Ap	0–20	Very dark brown (10YR 2/2) moist color, clayey loam texture, subangular blocky structures, pH 6.7, organic matter 2.5%
A	20–40	Dark brown (10YR 3/3) moist color, yellowish brown (10YR 5/6) mottling moist color, loam texture, subangular blocky structure, pH 7.5, organic matter 1.9%
AC	40–65	Dark grayish brown (10YR 4/2) moist color, yellowish red (5YR 4/6) mottling moist color, sandy loam texture, subangular blocky to porous massive structure, pH 7.9, organic matter 1.5%
CA	65–100	Primary color grayish brown (10YR 5/2) primary moist color, olive brown (2.5Y 4/3) secondary moist color, strong brown (7.5YR 4/6) mottling primary moist color, olive (5Y 4/3) mottling secondary moist color, sandy loam texture, porous massive to subangular blocky structure, pH 8.5, organic matter 0.5%
C	100–130	Dark yellowish brown (10YR 3/6) moist color, sandy texture, single grain massive structure, pH 8, organic matter 0.2%

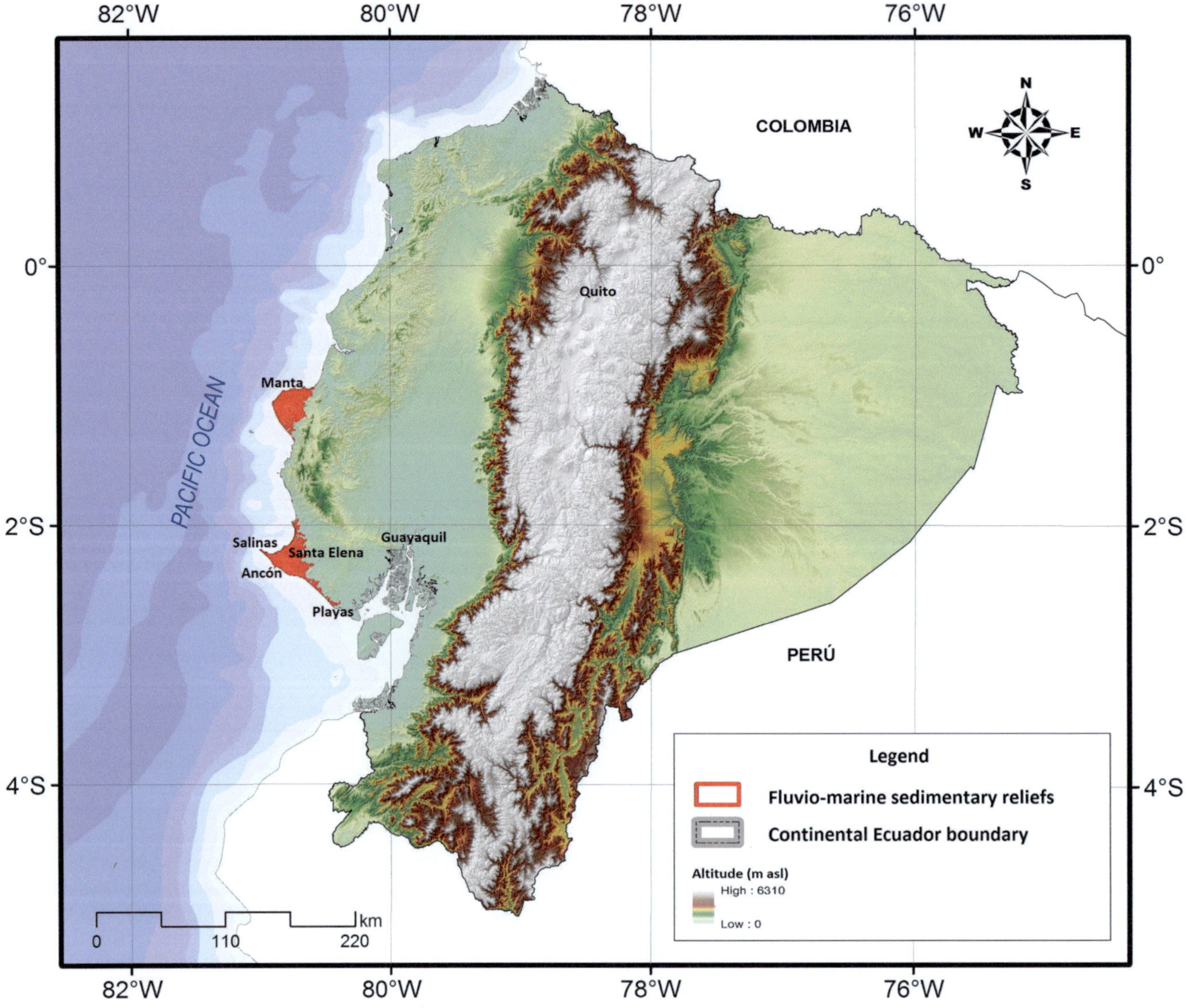

Fig. 2.33 Location of sedimentary and fluvio-marine reliefs with respect to continental Ecuador (adapted from Winckell and Zebrowski 1997)

sequence: A/BAk/Bkk/Ckm. Colors of the upper horizons are very dark grayish brown to very pale brown, while the Ckm horizon is white (IEE 2015).

Low Undulations from the More Recent Plains

This zone consists of reliefs formed by not consolidated Quaternary marine deposits of fine and coarse limestone sands containing a high percentage of gravel and shells fragments. These deposits are alternated with clayey sands and clays layers of cream to greenish color and conglomeratic sandstone beds rich in shells. These landscapes also belong to the Tablazo formation and cover almost totally the Salinas Peninsula and continue down to Playas (Collot et al. 2009; Winckell and Zebrowski 1997; Zebrowski and Sourdat 1997).

The profile of a representative soil from this zone is shown in Fig. 2.36 and Table 2.27. This soil, located on gentle slopes (<12%) and classified as Typic Calciargids (Soil Survey Staff 2006), has a salt excess throughout the

Fig. 2.34 Typic Haplustalfs soil profile (left) located at Las Palmas, Montecristi, Manabí, in a Marine Plateau (right) (IEE 2015)

Table 2.25 Profile characteristics of a soil classified as Typic Haplustalfs (IEE 2015)

Horizons	Depth (cm)	Description
A	0–15	Main color grayish brown (10YR 5/2) main dry color, loamy texture, subangular blocky structure, pH 7.1, organic matter 4.6%, CEC 24 cmol kg^{-1}, base saturation 93.6%
AE	15–40	Light brownish gray (10YR 6/2) dry color, silty loam texture, subangular blocky structure, pH 7, organic matter 1.3%, CEC 24 cmol kg^{-1}, base saturation 94.8%
Bt	40–106	Dark grayish brown (10YR 4/2) main dry color, clayey texture, prismatic structure, pH 7, organic matter 0.9%, CEC 30 cmol kg^{-1}, base saturation 90.77%
C	106–115+	Very pale brown (10YR 7/4) dry color, clayey loam texture, massive structure

Fig. 2.35 Modal profile of a soil classified as Calcic Petrocalcids (left) located at Cerro Barbasco, Santa Elena, at hills of tertiary origin (right) (IEE 2015)

Table 2.26 Profile characteristics of a soil classified as Calcic Petrocalcids (IEE 2015)

Horizons	Depth (cm)	Description
A	0–20	Grayish brown (10YR 5/2) dry color, very dark grayish brown (10YR 3/2) moist color, sandy-clayey loam texture, subangular blocky structure, light reaction to HCl, presence of secondary carbonates in the form of dispersed powdery lime, pH 8.2, organic matter 2.2%, CEC 30 cmol kg^{-1}, base saturation 95.3%
BAk	15–40	Very pale brown (10YR 8/2) main dry color, light brownish gray (10YR 6/2) secondary dry color, very pale brown (10YR 7/3) main moist color, dark grayish brown (10YR 4/2) secondary moist color, clayey loam texture, massive to subangular blocky structure, moderate reaction to HCl, presence of secondary carbonates as dispersed powdery lime
Bkk	40–106	Very pale brown (10YR 8/3) dry color, very pale brown (10YR 7/3) moist color, sandy loam texture, extremely strong reaction to HCl, presence of secondary carbonates as dispersed powdery lime
Ckm	106–115+	White (10YR 8/1) dry color, loamy sand texture, massive structure, extremely strong reaction to HCl, cemented layer by carbonates compaction in more than 90% of the horizon

profile, condition which limits crop production. Morphologically, the horizon sequence is A/Bt/Ck (IEE 2015).

Littoral Deposits

This zone corresponds to quaternary sandy marine deposits and calcareous saline deposits of silts and clays that also contain abundant shell residues that are located on beaches, beach ridges, and filled lagoons behind these strings. Beaches and coastal strings are composed of fine and very fine sand, often mobilized by wind. The interior lagoons are commonly composed by clayey to silty clayey deposits. The eolian process acting upon the littoral outline, especially on the south coast, promoted by the strong ocean winds, remobilize the current and old sand deposits forming dunes (Winckell and Zebrowski 1997; Zebrowski and Sourdat 1997).

The soils of these emerged beaches and coast strings are classified as Typic Ustipsamments (Fig. 2.37; Table 2.28) (Soil Survey Staff 2006). They show little or no evidence of

Fig. 2.36 Profile of a soil profile classified as Typic Calciargids (left) located at the Fuerte Militar, Salinas, Santa Elena, in a low undulation from the more recent plains. Note the very dry vegetation (right) (IEE 2015)

Table 2.27 Profile characteristics of a soil classified as Typic Calciargids (IEE 2015)

Horizons	Depth (cm)	Description
A	0–30	Pale yellow (2.5Y 7/3) dry color, brown (10YR 4/3) moist color, sandy loam texture, pH 7.8, organic matter 1.6%, CEC 18 cmol kg^{-1}, base saturation 91.3%
Bt	30–60	Dark yellowish brown (10YR 4/6) dry color, dark yellowish (10YR 4/4) moist color, sandy-clayey loam texture, prismatic structure with many clay liners, pH 8, organic matter 0.4%, CEC of 20 cmol kg^{-1}, base saturation 98.1%
Ck	60–110	White (10YR 8/1) dry color, pale yellow (2.5Y 8/2) moist color, sandy loam texture, porous massive structure, extremely strong reaction to HCl, pH 8.7, organic matter 0.4%, CEC of 30 cmol kg^{-1}, of base saturation 96.5%

pedogenic development, have an Ustic moisture regime (dry soil 90 or more cumulative days annually), and are located on gentle slopes (5–12%). Also, these soils have sandy texture, and consequently the excessive drainage (IEE 2015).

2.9 El Progreso Basin

This is a zone of subsidence located in the triangle formed by the Chanduy–Playas mountain range to the west and southwest and coastal Chongón and Colonche mountain range to the north. This unit occupies the eastern half of the Salinas Peninsula, including the Puna Island located immediately southeast of the peninsula, at the axis of the Guayas River delta, an island that maintains the continuity of the major structural features from the basin (Fig. 2.38). Overall, this basin is formed by reliefs of sedimentary-structural origin from the Tertiary period around a northwest southeast axis that has been filled by recent Oligo-Miocene sediments as it moves eastward into the basin. Altitude varies from 50 to 200 m o.s.l. Landscape ranges from flat to slightly wavy at the eastern boundary, and vigorous reliefs and traces of monocline reliefs

Fig. 2.37 Profile of a soil classified as Typic Ustipsamments located at Los Arenales, Crucita, Manabí (IEE 2015)

Table 2.28 Profile characteristics of a soil classified as Typic Ustipsamments (IEE 2015)

Horizons	Depth (cm)	Description
AC	0–20	Light brownish gray (10YR 6/2) dry color, dark grayish brown (10YR 4/2) moist color, sandy texture, unstructured, pH 9.0, organic matter 0.29%, CEC 30 cmol kg^{-1}, of base saturations 94.7%, moderate reaction to HCl
C1	20–50	Light brownish gray (10YR 6/2) dry color, grayish brown (10YR 5/2) moist color, sandy texture, unstructured, moderate reaction to HCl
C2	50–100	Light brownish gray (10YR 6/2) dry color, brown (10YR 5/3) moist color, sandy texture, unstructured, moderate reaction to HCl

(slopes and plateaus) of clusters of sandstone and clays, as well as clay-gravel depressions, all located to the west (MAG and ORSTOM 1978; Winckell and Zebrowski 1997).

The climate of El Progreso Basin is characterized by average annual rainfall ranging from 400 mm at the west to 1000 mm at the east. Similarly, rainy days change from 20 to 80, the number of dry months from 12 to 9, and the annual water deficit from 1000 to 750 mm in west to east direction. The mean annual temperature is around 22 °C. Considering the above, the following climatic regimes can be differentiated in this zone: (a) arid west and southwest; (b) very dry inner crown; (c) dry northeast; and (d) wet high reliefs covered by drizzle clouds. The Puna island climate is very dry (CLIRSEN et al. 2011a; IEE 2015; MAG and ORSTOM 1978; Winckell and Zebrowski 1997).

The soils in this unit are influenced by three formation factors: (a) Dry climate, predominant factor that affected physical and chemical pedogenic reactions favoring the synthesis of 2:1 clays, base accumulation, and frequent presence of carbonates and gypsum deposits; (b) Parental material that influenced the type of minerals inherited by soil, for example, the high K^+ level in these soils is due to

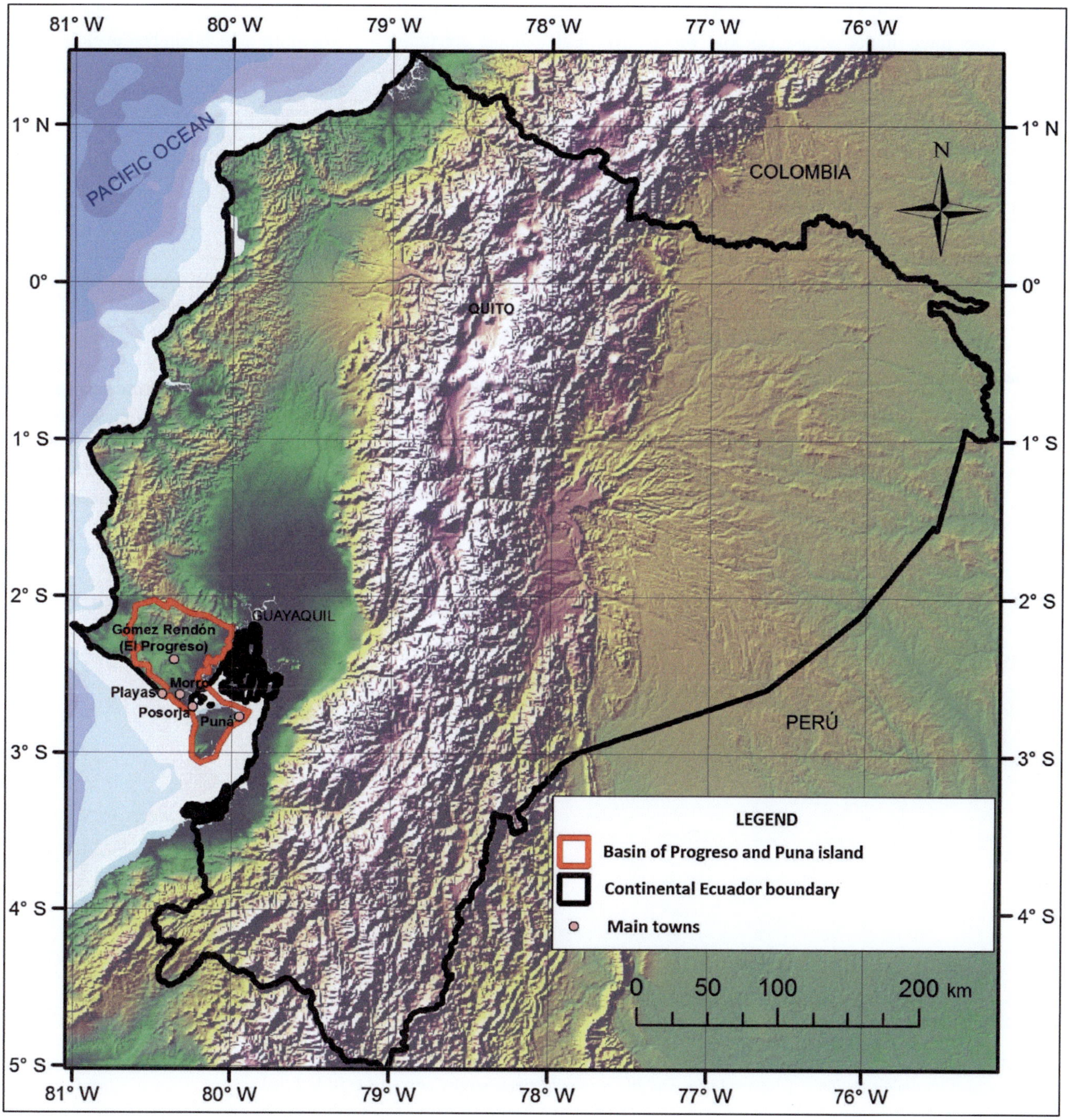

Fig. 2.38 Location of El Progreso Basin in relation to continental Ecuador (adapted from Winckell and Zebrowski 1997)

the presence of feldspars and micaceous primary rocks; and (c) Relief, soils are younger as more rugged is the relief because these areas are exposed to constant erosion (MAG and ORSTOM 1978; Mejía 1997).

Historically, El Progreso basin, also known as Gómez Rendón basin, has not been an area of significant agricultural exploitation due to climate constraints, particularly low rainfall. Common food crops like corn and peanuts were grown around the villages, and extensive livestock, especially goats, used the areas of natural vegetation for grazing purposes (Winckell and Zebrowski 1997). However, with the building of the hydraulic structure Chongón–San Vicente to transfer water to this dry area, associated small farmers are producing crops like corn, melon, watermelon, beans, and pepper on areas >4000 ha (MAGAP 2014). This unit is divided into three major areas: (a) low reliefs; (b) high reliefs; and (c) large dispersed glacis.

2.9.1 Low Reliefs

This is a zone of monotonous to gently wavy plains reliefs with wide slightly marked round plateaus sitting over clayey silty and clayey materials which formed soils classified as Vertisols (MAG and ORSTOM 1978; Winckell and Zebrowski 1997).

The modal profile of this soil, classified as Aridic Haplusterts (Soil Survey Staff 2006) (Fig. 2.39; Table 2.29), is characterized by cracks, slickensides, and high 2:1 clay content. The effective depth is only 35 cm. The pH is slightly <7.0 in the first horizon, but increases with depth. The presence of gypsum in the profile is inherited from the original parent materials (IEE 2015; MAG and ORSTOM 1978; Mejía 1997; Winckell et al. 1997).

These are fertile soils, but the lack of water has been the major constraint for agriculture which can be developed with the use of irrigation (Mejía 1997). Currently, the availability of water delivered by the Chongón–San Vicente hydraulic structure has allowed the use of this zone with high-production agriculture.

2.9.2 High Reliefs

The high reliefs zone was constituted over sandstones and conglomerates with altitudes not exceeding 275 m o.s.l. (Winckell and Zebrowski 1997). Soils are poorly developed, since the erosion processes are superior to the pedogenic processes and, therefore, they are characterized by a shallow profile which leaves little space for root development as observed in the modal profile presented in Fig. 2.40 and Table 2.30. This soil, classified as Lithic Ustorthents (Soil Survey Staff 2006), has slightly weathered rock strata within the first 50 cm of the profile. In all horizons, there is medium

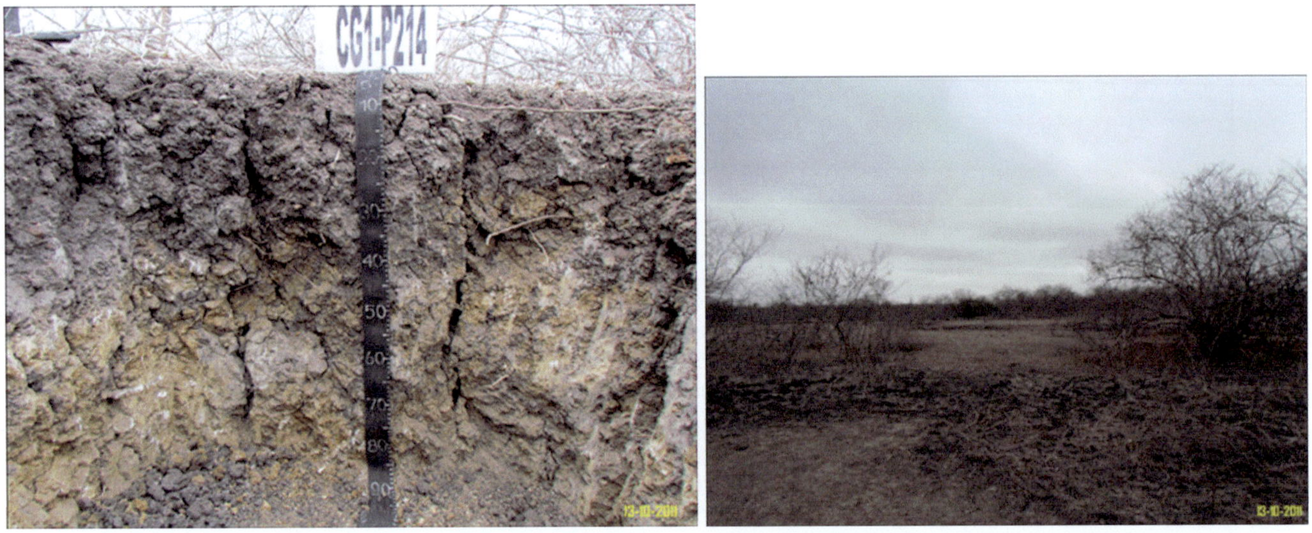

Fig. 2.39 Profile of a soil classified as Aridic Haplusterts (left), located at El Ayalán, El Morro, Guayas, with a the following horizon sequence: A/CAy/Cy over an undulating relief (slope < 12%). The site is at a depopulated dry area (right) (IEE 2015)

Table 2.29 Profile characteristics of a soil classified as Aridic Haplusterts (IEE 2015)

Horizons	Depth (cm)	Description
A	0–35	Very dark gray (10YR 3/1) dry color, very dark gray (2.5Y 3/1) moist color, clayey loam texture, subangular prismatic structure, presence of coarse fragments of gravel, pH 6.4, organic matter 0.4%, CEC 37 cmol kg^{-1}, base saturation 98.1%
CAy	35–75	Clear yellowish brown (2.5Y 6/3) main moist color, light olive brown (2.5Y 5/6) secondary moist color, clayey texture, massive structure, presence of coarse fragments of gravel
Cy	75–85	Light olive brown (2.5Y 5/4) moist color, clayey texture, massive structure, presence of gypsum crystals

Fig. 2.40 Profile of a soil classified as Lithic Ustorthents (left) with the following horizon sequence: Ak/Crk1/Crk2/R, located at Pampas de Calicante, Santa Elena. The site is located in a medium wavy undulating relief (right) (IEE 2015)

Table 2.30 Profile characteristics of a soil classified as Lithic Ustorthents (IEE 2015)

Horizons	Depth (cm)	Description
Ak	0–5	Light olive brown (2.5Y 5/4) dry color, light olive brown (2.5Y 5/4) moist color, loamy texture, angular blocky structure, moderate reaction to HCl, abundant coarse fragments of fine gravel, many coatings of calcium carbonate, pH 8.5, organic matter 0.3%, CEC 29 cmol kg^{-1}, base saturation 97.2%
Crk1	5–10	Light olive brown (2.5Y 5/4) main dry color, yellowish brown (10YR 5/6) secondary dry color; light olive brown (2.5Y 5/4) main moist color, yellowish brown (10YR 5/6) secondary moist color, silty clay texture, massive structure, abundant coarse gravel fragments, moderate reaction to HCl, powdery lime, many coatings of calcium carbonate
Crk2	10–45	Pale yellow (2.5Y 7/3) dry color, light olive brown (2.5Y 5/4) moist color, loamy sand texture, massive structure, dominant presence of coarse fragments of coarse gravel, strong reaction to HCl, powdery lime, few calcium carbonate coatings
R	45+	Rock

to strong reaction to HCl due to the presence of calcium carbonate (MIDENA et al. 2012a). These soils with a shallow rocky horizon are located on steep slopes susceptible to erosion; for these reasons, these soils are not used for agriculture.

2.9.3 Large Dispersed Glacis

This zone corresponds to very dissected glacis located mainly around the large Chanduy–Playas hill, site that presents a fairly regular and continuous crown which progressively descends from around 100 m o.s.l. down to the point it gets in contact with Low Reliefs. These glacis come from scattered colluvial-alluvial materials on which stony fersiallitic soils have developed over sandy-clayey reddish matrix having a strong secondary ferruginous cementation that should have been developed in different weather conditions that those prevalent today, conditions which are not conductive for movement and crystallization of the metal oxides (MAG and ORSTOM 1978; Winckell and Zebrowski 1997).

The representative profile of the soils of this zone is presented in Fig. 2.41 and Table 2.31. This soil is classified as Arenic Paleargids (Soil Survey Staff 2006) and is considered a Paleosol because its current characteristics do not

Fig. 2.41 Profile of a soil classified as Arenic Paleargids (left), located at Cerro El Morro, Guayas with the following horizon sequence: A/Bt/C. The site at a dispersed glacis landscape (right) has surface gravel sandstone and cherts (IEE 2015)

Table 2.31 Characteristics of a soil profile classified as Arenic Paleargids (IEE 2015)

Horizons	Depth (cm)	Description
A	0–10	Yellowish red (5YR 5/6) main dry color, pale brown (10YR 6/3) secondary dry color, reddish brown (5YR 4/4) main moist color, brown (10YR 5/3) secondary moist color, sandy loam texture, subangular blocky structure, many coarse gravel fragments, pH 5.1, organic matter 1%, CEC 9.0 cmol kg^{-1}, base saturation 90%
Bt	10–25	Brown (10YR 5/3) dry color, pale brown (10YR 6/3) moist color, sandy clay loam texture, prismatic structure, many coarse gravel fragments, pH 5.5, organic matter 0.4%, CEC 21 cmol kg^{-1}, base saturation 94.4%
C1	25–45	Pale brown (10YR 6/3) main dry color, brown (10YR 5/3) secondary dry color, very pale brown (10YR 7/3) main moist color, light gray (10YR 7/2) secondary moist color, sandy loam texture, massive structure, with coarse gravel fragments
C2	45–60	Pale yellow (2.5Y 8/4) dry color, pale yellow (2.5Y 7/3) moist color, loamy sand texture, massive structure, many coarse gravel fragments

correspond to the present-day climate. The effective depth is only 10 cm. According to the weighted average of the first two horizons, there is a medium level of exchangeable acidity (1.23 cmol kg^{-1}), low organic matter (0.64%), medium CIC (16.2 cmol kg^{-1}), and a high percentage base saturation (92.6%) (IEE 2015). Agricultural production is almost impossible in these soils due to the very dry climate and the presence of abundant coarse gravel (2–2.5 cm) on the surface which prevents mechanization.

2.10 Chongón Colonche Mountain Range

The Chongón Colonche mountain range, with an arch structure that extends from El Membrillal to Guayaquil (Fig. 2.42), is the most important and imposing massif of the coastal plain. The highest point reaches an altitude of 830 m o.s.l. The highest reliefs (dissected massifs) are sited over volcanic and volcano-sedimentary outcrops, while monoclines reliefs, more or less upright, developed over sedimentary strata associated with steep and highly dissected areas. This explains the marked asymmetry of this cordillera; the southern flank is quite dissected and has a steep slope, while the northeast side has a succession of gently sloping monocline reliefs (Winckell and Zebrowski 1997).

The altitudinal gradient strongly influences the climate of this unit. The annual rainfall in the lowlands is <500 mm (10–12 dry months); while in the highlands rainfall is around 1000 mm (4–8 dry months). There is also a strong contrast between the two ends of the cordillera; in northern horn, closer to the seashore, conditions are drier (400 mm annual rainfall), while in the southern horn, precipitation is close to 1000 mm. The zone has an average annual temperature of 25 °C, with a difference of 3 °C among extreme monthly mean temperatures (Winckell and Zebrowski 1997).

There is no significant agricultural production in this unit mainly due to steep reliefs, compact shaping, the presence of shallow soils, and severe drought conditions. Dry zones are

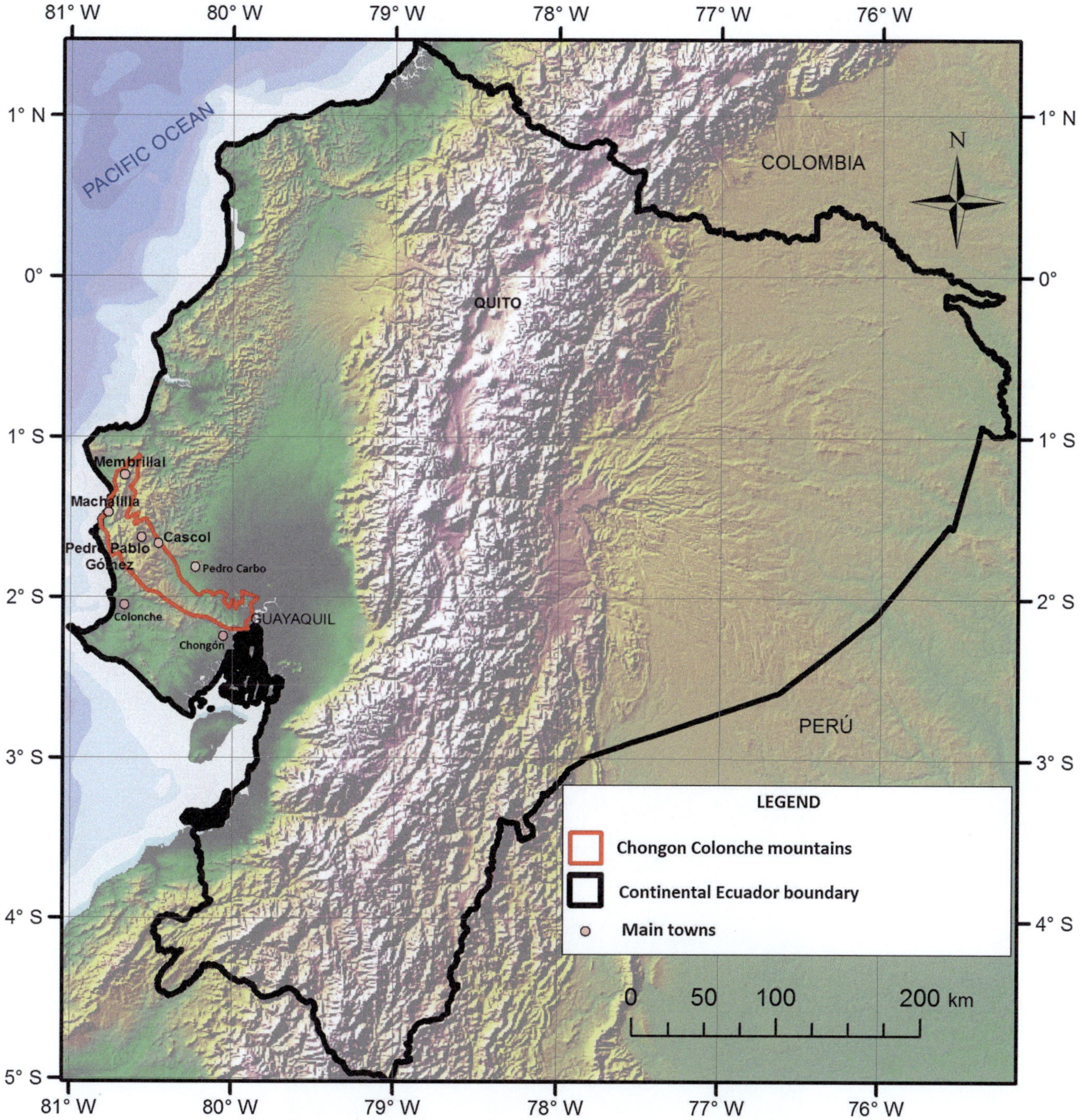

Fig. 2.42 Location of the Chongón Colonche mountain range with respect to continental Ecuador (adapted from Winckell and Zebrowski 1997)

used for itinerant cattle, goats, and sheep grazing. The highest and more humid areas are used for marginal citrus and coffee production. Agricultural activity on the flat and moderately shaped areas runs around crops such as coffee, banana, cocoa, and citrus, in many cases associated with cattle production (Winckell and Zebrowski 1997). However, more intense production of cocoa and coffee, through communal farmers association, has developed in this zone (ACBIO 2012; Bonifaz et al. 2004).

This diversity of reliefs and the particular arrangement of the different landscapes have shaped four major sets of conditions for soil development in this unit: (a) shallow reliefs, usually on rock; (b) high reliefs; (c) low reliefs; and (d) alluvial environments.

2.10.1 Shallow Reliefs with Rock Outcroppings

The soils in this zone were developed over ancient volcanic rocks (basic lavas, tuffs, breccias) which reappeared as a consequence of major tectonic movements that caused significant flaws in the relief at the heart of the cordillera. These rocks have high resistance to weathering, and are located in dry and very dry areas; for these reasons, only shallow alterations were developed forming clayey–stony soils with numerous boulder outcroppings on the surface (Winckell and Zebrowski 1997; Zebrowski and Sourdat 1997).

A characteristic profile belongs to a soil classified as Lithic Haplusterts (Soil Survey Staff 2006) which is shown in Fig. 2.43 and Table 2.32. The soil is located at an area with general 40–70% slope with Ustic and isohyperthermic moisture and soil temperature regimes, respectively. Morphologically, the profile has the following horizon sequence: A/AC/R. The surface texture is clayey loam with the presence of bedrock at 39 cm down the surface. The soil is characterized by deep cracks and slickensides related to the high content of 2:1 clay (montmorillonite) (IEE 2015).

2.10.2 High Reliefs

These are the most impressive and highest reliefs of the cordillera, derived from a sedimentary-volcanic sandstone complex, graywacke, and interbedded siliceous facies of lava and pillow lava (Cayo formation), ending at the top with a series of silicified rocks and cherts (Guayaquil formation). The central massifs, having the highest points of the cordillera (between 500 and 800 m o.s.l.), are characterized by large rectilinear flanks, steep to very steep slopes (>70%) at the north, while the full southwestern flank consists of vigorous reliefs with altitudes rising from 200 to 830 m o.s.l. These reliefs are the mountain front which blocks atmospheric circulation causing orographic precipitations that have altered the rock forming soils with mollic

Fig. 2.43 Profile of a soil profile classified as Lithic Haplusterts, located at Estero Verde, Isidro Ayora, Guayas (IEE 2015)

characteristics adjoining rocky and stony outcroppings (Winckell and Zebrowski 1997; Zebrowski and Sourdat 1997). Due to the described climatic conditions, the modal profile of this zone corresponds to a soil classified as Entic Hapludolls (Soil Survey Staff 2006) (Fig. 2.44; Table 2.33). This shallow soil, located on a steep slope, presents the following horizon sequence: Ap/A/BC/C/Cr (IEE 2015).

2.10.3 Low Reliefs

The reliefs are made of very soft hills with relative gradient from 15 to 25 m, formed from more or less cemented siltstone and limonite, extending in a halo of reliefs from 40 to 80 m o.s.l. in the site adjoin the alluvial plain and then gradually rising up to 80–100 m o.s.l. north of Pedro Carbo

Table 2.32 Profile characteristics of a soil classified as Lithic Haplusterts (IEE 2015)

Horizons	Depth (cm)	Description
A	0–28	Very dark gray (2.5Y 3/1) dry color, very dark gray (10YR 3/1) with red yellowish mottling (5YR 5/8) moist color, clayey loam texture, angular blocky structure, abundant coarse fragments of gravel, pH 6.6, organic matter 3.5%, CEC 37.0 cmol kg^{-1}, base saturation 94.1%
AC	28–39	Very dark grayish brown (10YR 3/2) dry color, dark reddish brown mottling (5YR 3/4) moist color, clayey texture, angular blocky structure, many coarse gravel fragments, pH 6.6, organic matter 0.7%, CEC 47.0 cmol kg^{-1}, base saturation 79.2%
R	29–60+	Rock (lutite)

Fig. 2.44 Profile of a soil classified as Entic Hapludolls (left) located at Río Blanco, Puerto Lopez, Manabí. The site shows a high relief with steep slopes (right) (IEE 2015)

Table 2.33 Profile characteristics of a soil classified as Entic Hapludolls (IEE 2015)

Horizons	Depth (cm)	Description
Ap	0–10	Very dark grayish brown (10YR 3/2) moist color, loamy texture, granular structure, pH 6.9, organic matter 2.9%, CEC 56 cmol kg^{-1}, base saturation 97.1%
A	10–35	Brown (10YR 4/3) dry color, very dark grayish brown (10YR 3/2) moist color, loamy texture, angular and subangular blocky structure, pH 6.9, organic matter 2%, CEC of 57 cmol kg^{-1}, base saturation 97.5%
BC	35–50	Main color brown (10YR 4/3) main dry color, dark grayish brown (10YR 4/2) secondary dry color, very dark grayish brown (10YR 3/2) moist color, clayey loam texture, massive to subangular blocky structure, pH 7.1, organic matter 0.6%
C	50–75	Yellowish brown (10YR 5/4) dry color, brown (10YR 4/3) moist color, sandy clay texture, massive structure, common presence of coarse gravel fragments
Cr	75–115	Light olive brown (2.5Y 5/6) main dry color, light olive brown (2.5Y 5/4) secondary dry color, olive brown (25Y 4/4) main moist color, olive brown (2.5Y 4/3) secondary moist color, sandy clay texture, massive structure

Fig. 2.45 Profile of a soil classified as Typic Haplusterts (left), located at La Providencia, Pedro Carbo, Guayas, in low hill relief (slope <12%) (right) (IEE 2015)

and 200 m o.s.l. at Cascol. These landscapes of depressed hills and ondulating landscapes, located beneath the Tertiary sedimentary reliefs of sandy slopes and conglomerates of the cordillera, have convex–concave smooth flanks and wide flat recessed tops with slopes ranging from 12 to 25%. Rock alterations on these soft monotonous reliefs with a relatively dry climate have allowed the development of clayey to silty clay soils of moderate depth (IEE 2015). The modal profile is from a soil classified as Typic Haplusterts (Soil Survey Staff 2006) (Fig. 2.45; Table 2.34), developed in an Ustic moisture and isohyperthermic temperature regimes. The horizon sequence is as follows: A1/A2/AC/C (IEE 2015).

Table 2.34 Profile characteristics of a soil classified as Typic Haplusterts (IEE 2015)

Horizons	Depth (cm)	Description
Ap	0–19	Gray (2.5Y 5/1) dry color, dark gray (10YR 4/1) with yellowish red mottling (5YR 5/8) moist color, sandy-clayey loam texture, angular blocky structure, coarse gravel fragments, pH 6.1, organic matter content 0.3%, CEC 24 cmol kg^{-1}, base saturation 96.8%
A	19–42	Gray (10YR 5/1) dry color, very dark gray (2.5Y 3/1) with yellowish brown mottling (10YR 5/8) moist color, sandy-clayey loam texture, angular blocky structure, many coarse gravel fragments, pH 6.2, organic matter 0.4%, CEC 27 cmol kg^{-1}, base saturation 95.8%
AC	42–61	Gray (7.5YR 5/1) dry color, brownish yellow (10YR 6/6) moist color, sandy-clayey loam texture, massive to angular blocky structure, pH 6.2, organic matter 0.7%, CEC 25 cmol kg^{-1}, base saturation 95.4%
C	61–95+	Light yellowish brown (2.5Y 6/4) moist color, sandy-clayey texture, massive structure

Fig. 2.46 Profile of a soil classified as Vertic Ustifluvents, located at El Paraiso, Pedro Carbo, Guayas (IEE 2015)

2.10.4 Alluvial Environments

The alluvial areas correspond to the fluvial valleys with complexes of terraces (low, medium, and high), located mainly between the bottom of the cordillera and the contact with the recent alluvial plain southeast of Pedro Carbo. Soils from this area have variable texture, usually silty upstream, and more clayey downstream, utilized for pasture, plantain, maize, and cassava, but later used for cocoa and coffee cultivation by the farmer communities of the region. A modal profile belongs to a soil classified as Vertic Ustifluvents (Soil Survey Staff 2006) (Fig. 2.46; Table 2.35), located at a medium terrace of alluvial deposits with slopes from 2 to 5%. This soil has Ustic and isohyperthermic moisture and temperature regimes, respectively. Morphologically, this soil presents a profile with the following horizon sequence: A/C1/2C2/2C3/3C4 (IEE 2015).

Table 2.35 Profile characteristics of a soil classified as Vertic Ustifluvents (IEE 2015)

Horizons	Depth (cm)	Description
A	0–43	Dark gray (10YR 4/1) dry color, black (10YR 2/1) moist color, clayey loam texture, subangular blocky structure, pH 7.3, organic matter 2.5%, CEC 28 cmol kg^{-1}, base saturation 95.3%
C1	43–78	Light olive brown (2.5Y 5/4) moist color, sandy loam texture, massive structure, pH 7.6, organic matter 0.7%
2C2	78–95	Olive brown (2.5Y 4/4) main moist color, very dark gray (2.5Y 3/1) secondary moist color, loamy texture, pH 7.3, organic matter 1%
2C3	95–107	Very dark gray (2.5Y 3/1) moist color, light olive brown (2.5Y 5/4) secondary moist color, clayey loam texture, pH 7.6, organic matter 0.9%
3C4	107–145+	Light yellowish brown (2.5Y 6/3) moist color, sandy loam texture, massive structure, pH 7.3, organic matter 0.3%

References

Aalto KR, Miller Iii W (1999) Sedimentology of the Pliocene Upper Onzole Formation, an inner-trench slope succession in northwestern Ecuador. J S Am Earth Sci 12(1):69–85

ACBIO (2012) Plan de acción del biocorredor Chongón Colonche. Biocorredores para el Buen Vivir. FIDES, SGP, GEF, UNDP, Guayaquil, Ecuador

Bonifaz C, Elao R, Santos M, Avilés J, Cornejo X (2004) Plan general de manejo del bosque protector de la cordillera Chongón Colonche. Universidad de Guayaquil, Guayaquil, Ecuador

CLIRSEN, SENACYT, MAGAP, SIGAGRO (2009a) Memoria técnica de suelos del cantón Colimes. Proyecto: Generación de geoinformación para la gestión del territorio y valoración de tierras rurales en la cuenca del río Guayas escala 1: 25 000. CLIRSEN, Quito, Ecuador

CLIRSEN, SENACYT, SENPLADES, MAGAP, SIGAGRO (2009b) Memoria técnica de la geopedología del cantón Naranjito. Proyecto: Generación de geoinformación para la gestión del territorio y valoración de tierras rurales en la cuenca del río Guayas escala 1: 25 000. CLIRSEN, Quito, Ecuador

CLIRSEN, SENACYT, SENPLADES, MAGAP, SIGAGRO (2009c) Memoria técnica de suelos del cantón Samborondón. Proyecto: Generación de geoinformación para la gestión del territorio y valoración de tierras rurales en la cuenca del río Guayas escala 1: 25 000. CLIRSEN, Quito, Ecuador

CLIRSEN, SENACYT, SENPLADES, MAGAP, SIGAGRO (2009d) Memoria técnica de suelos del cantón Vinces. Proyecto: Generación de geoinformación para la gestión del territorio y valoración de tierras rurales en la cuenca del río Guayas escala 1: 25 000. CLIRSEN, Quito, Ecuador

CLIRSEN, SENPLADES, MAGAP, INIGEMM (2011a) Memoria técnica de geopedología del cantón Montecristi. Proyecto: Generación de geoinformación para la gestión del territorio a nivel nacional escala 1: 25 000. CLIRSEN, Quito, Ecuador

CLIRSEN, SENPLADES, MAGAP, INIGEMM (2012a) Memoria técnica de la geopedología del cantón Santa Ana. Proyecto: Generación de geoinformación para la gestión del territorio a nivel nacional escala 1: 25 000. CLIRSEN, Quito, Ecuador

Collot J-Y, Michaud F, Alvarado A, Marcaillou B, Sosson M, Ratzov G, Migeon S, Calahorrano A, Pazmino A (2009) Visión general de la morfología submarina del margen convergente de Ecuador-Sur de Colombia: implicaciones sobre la transferencia de masa y la edad de la subducción de la Cordillera de Carnegie. In: Comisión Nacional del Derecho del Mar (CNDM) (ed) Geología y geofísica marina y terrestre del Ecuador desde la costa continental hasta las islas Galápagos. CNDM-IRD-INOCAR, Guayaquil, Ecuador

Huttel C (1997) Las grandes regiones climáticas y sus formaciones vegetales naturales. In: Winckell A, Marocco R, Winter T et al (eds) Paisajes naturales del Ecuador: las condiciones generales del medio natural. CEDIG - IPGH - IRD - IGM, Quito

Huttel C, Zebronski C, Gondard P, Bourliaud J (1999) Paisajes agrarios de Ecuador., vol V. Geografía Básica del Ecuador. Instituto Panamericano de Geografía e Historia, Quito, Ecuador

IEE (2012a) Geovisualizador Instituto Espacial Ecuatoriano. Mosaico de Geopedología 1:25 000. IEE. http://www.institutoespacial.gob.ec/geoportal/601-2/. Accessed 18 de diciembre 2014

IEE (2012b) Geovisualizador Instituto Espacial Ecuatoriano. Mosaico de Uso de la tierra 1:25 000. IEE. http://www.institutoespacial.gob.ec/geoportal/601-2/. Accessed 18 de diciembre 2014

IEE (2015) Perfiles de suelos de Ecuador. IEE. http://www.institutoespacial.gob.ec/geoportal/perfiles-de-suelos/. Accessed 10 de Octubre 2015

INEC (2013) Encuesta de superficie y producción agropecuaria continua Ecuador 2013. Dirección de Estadísticas Agropecuarias y Ambientales, INEC, Quito, Ecuador

Lasso L, Cruz G, Haro R (2010) Zonificación agroecológica de tres cultivos estratégicos: maíz (*Zea mays* L.), arroz (*Oryza sativa* L.) y caña de azúcar (*Saccharum oficinarum* L.) en tres cantones de la cuenca baja del río Guayas. Paper presented at the XII Congreso Ecuatoriano de la Ciencia del Suelo: El suelo como factor de mitigación del cambio climático, Santo Domingo, Ecuador

León J (2010) Manual de geografia del Ecuador. Universidad Andina Simón Bolívar, sede Ecuador - Corporación Editora Nacional, Quito, Ecuador

MAG, ORSTOM (1978) Inventario catográfico de la Costa ecuatoriana. MAG-PRONAREG, Quito, Ecuador

MAGAP (2014) Proyecto PIDAASSE. MAGAP. http://www.agricultura.gob.ec/pidaasse/. Accessed 5 de febrero 2015

Mejía L (1997) Suelos del Ecuador: Reconocimento general en base a su capacidad-fertilidad y mapa general de clasificación por Capacidad-Fertilidad de los suelos del Ecuador: Una interpretación básica sobre las características de los suelos del Ecuador Fundación Forestal Juan Manuel Durini; Instituto de la Potasa y el Fósforo (INPOFOS) - Instituto Geográfico Militar (IGM) - Instituto panamericano de Geografía e Historia (IPGH) - Sociedad Ecuatoriana de la Ciencia del Suelo (SECS), Quito, Ecuador

MIDENA, IEE, SEMPLADES, MAGAP, INIGEMM (2012a) Memoria técnica geopedológica del cantón Santa Elena. Proyecto: Generación de geoinformación para la gestión del territorio a nivel nacional escala 1: 25 000. IEE, Quito, Ecuador

MIDENA, IEE, SENPLADES, MAGAP (2013b) Memoria técnica geopedológica del cantón Valencia. Proyecto: Generación de geoinformación para la gestión del territorio a nivel nacional escala 1: 25 000. IEE, Quito, Ecuador

Ministerio de Transporte y Obras Públicas - INECA (2012) Estudio de ingeniería de la carretera Santo Domingo-Esmeraldas y la estructuración jurídica, técnica y económico-financiera. Geología y Geotecnia, vol 3. Ministerio de Transporte y Obras Públicas, Quito, Ecuador

Montúfar C, Yerovi F, Sánchez D, Lagos E (2010) Caracterización de la dinámica ambiental, a escala de tiempo humano, en el cantón milagro mediante el uso de herramientas geoespaciales. Paper presented at the Paper presented at the XII Congreso Ecuatoriano de la Ciencia del Suelo: El suelo como factor de mitigación del cambio climático, Santo Domingo, Ecuador

Moreno J (2001) Zonificación agroecológica de la cuenca baja del río Guayas-Ecuador con utilización de la Teledetección. Universidad Central del Ecuador, Quito, Ecuador

Pacheco R (2009) El Ecuador: Recursos naturales agrícolas y del medio ambiente. Imprenta Colón, Quito

PRONAREG-ORSTOM (1982) Inventario cartográfico de la Costa ecuatoriana: mapas de aptitudes agrícolas, escala 1: 200 000. Ministerio de Agricultura y Ganadería (MAG), Programa Nacional de Regionalización Agraria (PRONAREG), Office de la Recherche Scientifique et Technique Outre Mer (ORSTOM), Quito, Ecuador

Ramírez MF (2013) Registros de la deformación y del volcanismo en el dominio del antearco ecuatoriano: sedimentología y bioestratigrafía de la formación Borbón. Escuela Superior Politécnica del Litoral, Guayaquil Ecuador

Sevillano G (2010) Levantamiento geopedológico de la cuenca baja del río Guayas-Ecuador con aplicación del enfoque sistémico. Paper presented at the XII Congreso Ecuatoriano de la Ciencia del Suelo: El suelo como factor de mitigación del cambio climático, Santo Domingo, Ecuador

Soil Survey Staff (2006) Keys to soil taxonomy, 10th edn. USDA - NRCS, Washington

Tapia J (2012) Modelización hidrológica de un área experimental en la Cuenca del Río Guayas. Universidad Nacional de la Plata, Cuidad de la Plata, Argentina

Winckell A, Zebrowski C (1997) Los paisajes costeros. In: Winckell A, Zebrowski C, Sourdat M (eds) Los paisajes naturales del Ecuador: Las regiones y paisajes del Ecuador. CEDIG - IPGH - IRD, Quito

Winckell A, Zebrowski C, Sourdat M (1997) Los paisajes naturales del Ecuador, vol IV Las regiones y paisajes del Ecuador. Centro Ecuatoriano de Investigación Geográfica (CEDIG) - IPGH (Sección Ecuador) - IRD (Francia) - IGM, Quito, Ecuador

Zebrowski C, Sourdat M (1997) Los factores de la pedogénesis y los suelos en Ecuador. In: Winckell A, Marocco R, Winter T et al. (eds) Los paisajes naturales del Ecuador, vol Condiciones generales del medio natural. Centro Ecuatoriano de Investigación Geográfica (CEDIG) - IPGH (Sección Ecuador) - ORSTOM (Francia) - IGM, Quito, Ecuador

3 Soils from the Highlands

Julio Moreno, Fausto Yerovi, Mireya Herrera, Darwin Yánez, and José Espinosa

Abstract

The Andean Highlands are made of the western and eastern cordilleras arranged in a north–south direction, with very steep outer slopes characterized by a gradual decline in altitude as they diffuse into the coastal plain or the Amazonia, and a decreasing massiveness also from north to south. Highlands are divided into three regions: Northern, Central, and Southern Highlands. The northern part received recent contributions of volcanic projections (ash and lapilli) over which diverse fertile and generally deep Andisols have developed, characteristics closely linked to climate and parent material. The Central Highlands do not have active volcanoes or recent pyroclastic projections and parent material consists of old volcanic and metamorphic rocks in a monotonous relief. There are a variety of soils in this zone depending on the altitude. In the upper sectors, soils are ferralitic–fersalitic, in the middle, they are moderately evolved with mollic or vertic features, and in the lowland, the soils are poorly developed (Inceptisols and Entisols) . Southern Highlands are characterized by a more rugged relief over granites, metamorphic formations, tertiary sediments, and glacis of colluvial deposits that have formed, in the upper parts, fersiallitic saturated soils are rich in minerals 2:1, along with desaturated ferralitic Paleosols. In the lower areas, soils are mainly vertisoles, often poorly developed, surrounded by the fersiallitic Paleosols.

Keywords

Highlands • Soils of the north sierra • Soils of the central sierra • Soils of the south sierra

3.1 Introduction

The Andes or Highlands are the most notorious mountain barrier in Ecuador with a width varying from 100 and 150 km from east to west. The great Andes mountain range is comprised of the Western and Eastern Cordilleras arranged in meridian direction, which have very abrupt outer slopes characterized by a general decline in altitude as they diffuse into the coastal plain or the Amazonia, and by a decreasing massiveness from north to south.

The Highlands are divided into three regions (Fig. 3.1): (a) Northern Highlands, which stretches from the border with Colombia to latitude 2°15′S (Palmira-Alausí) and is characterized by the presence of several volcanoes, some of them

J. Moreno (✉) · F. Yerovi
Instituto Espacial Ecuatoriano, Quito, Ecuador
e-mail: julio.moreno@institutoespacial.gob.ec

F. Yerovi
e-mail: patoyerovi@gmail.com

M. Herrera
Ministerio de Agricultura y Ganadería (MAGAP), Quito, Ecuador
e-mail: mireyaherrera18@hotmail.com

D. Yánez
Secretaría de Gestión de Riesgos Ecuador, Guayaquil, Ecuador
e-mail: darwinbeniyanez@yahoo.es

J. Espinosa
Facultad de Ciencias Agrícolas, Universidad Central del Ecuador, Quito, Ecuador
e-mail: aespinosa@uce.edu.ec

J. Espinosa et al. (eds.), *The Soils of Ecuador*, World Soils Book Series,
https://doi.org/10.1007/978-3-319-25319-0_3

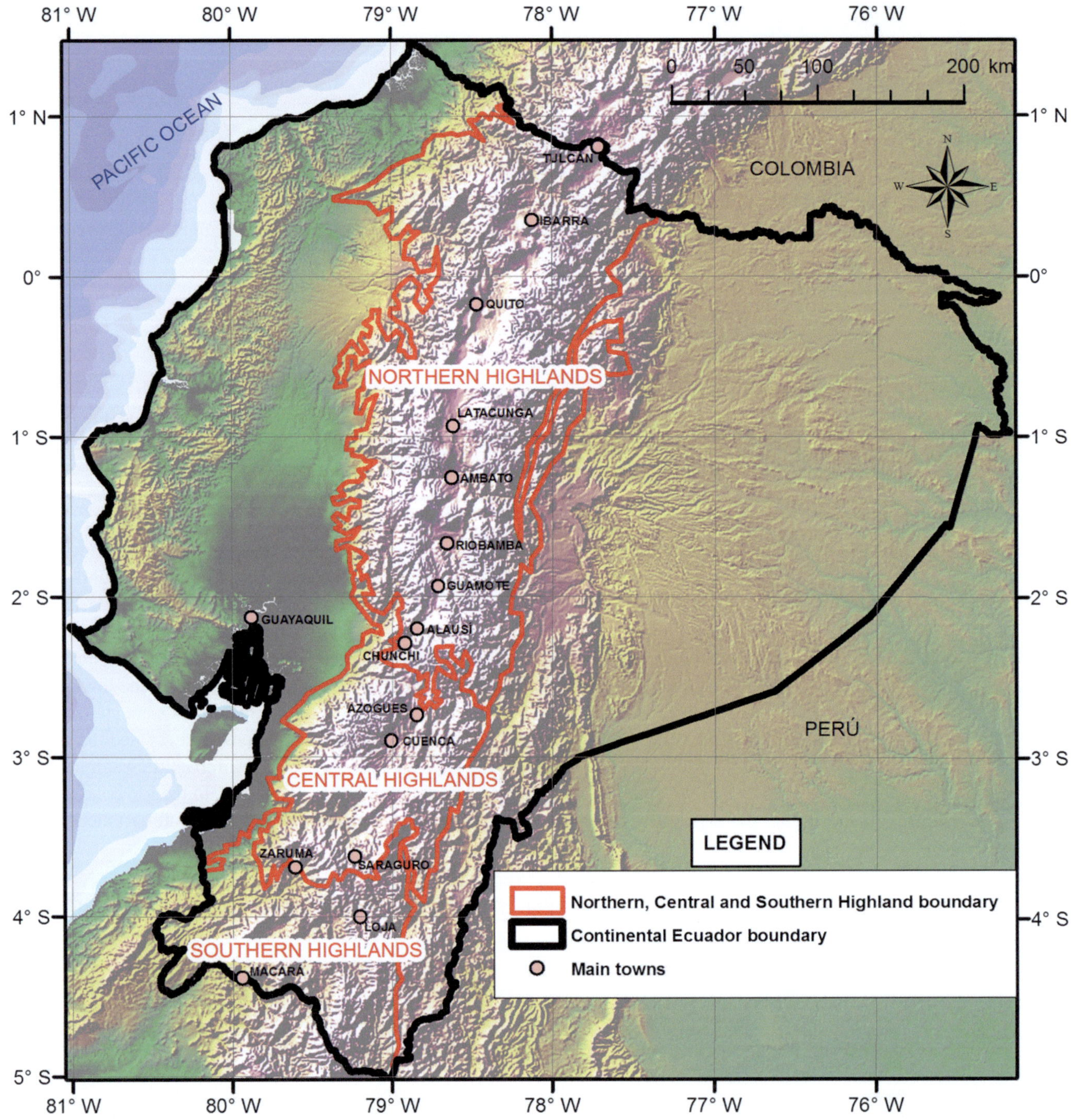

Fig. 3.1 Northern, Central, and Southern Highlands location with respect to continental Ecuador (Winckell et al. 1997)

active; (b) Central Highlands, extending from Palmira-Alausí to Zaruma-Saraguro at the south, without the presence of recent volcanoes and a landscape of a typical spread and monotonous plateau; and (c) Southern Highlands, which extends from Zaruma-Saraguro to the border with Peru, with a very irregular relief characterized by low areas that extend to the SSO and higher areas that barely reach 2500 m a.s.l. (IGM 2013; Winckell et al. 1997).

The characteristics of these three zones can be summarized in the following geosystems (Fig. 3.2): (a) Cold tops of the Andean cordilleras are characterized by a paramo landscape with cold to very cold climate and quotidian nighty frosts, presence of volcano cones of different ages, and poorly developed black soils with high organic matter content. Climate does not favor agriculture activity, but these areas are used for extensive sheep grazing; (b) Andes outer slopes,

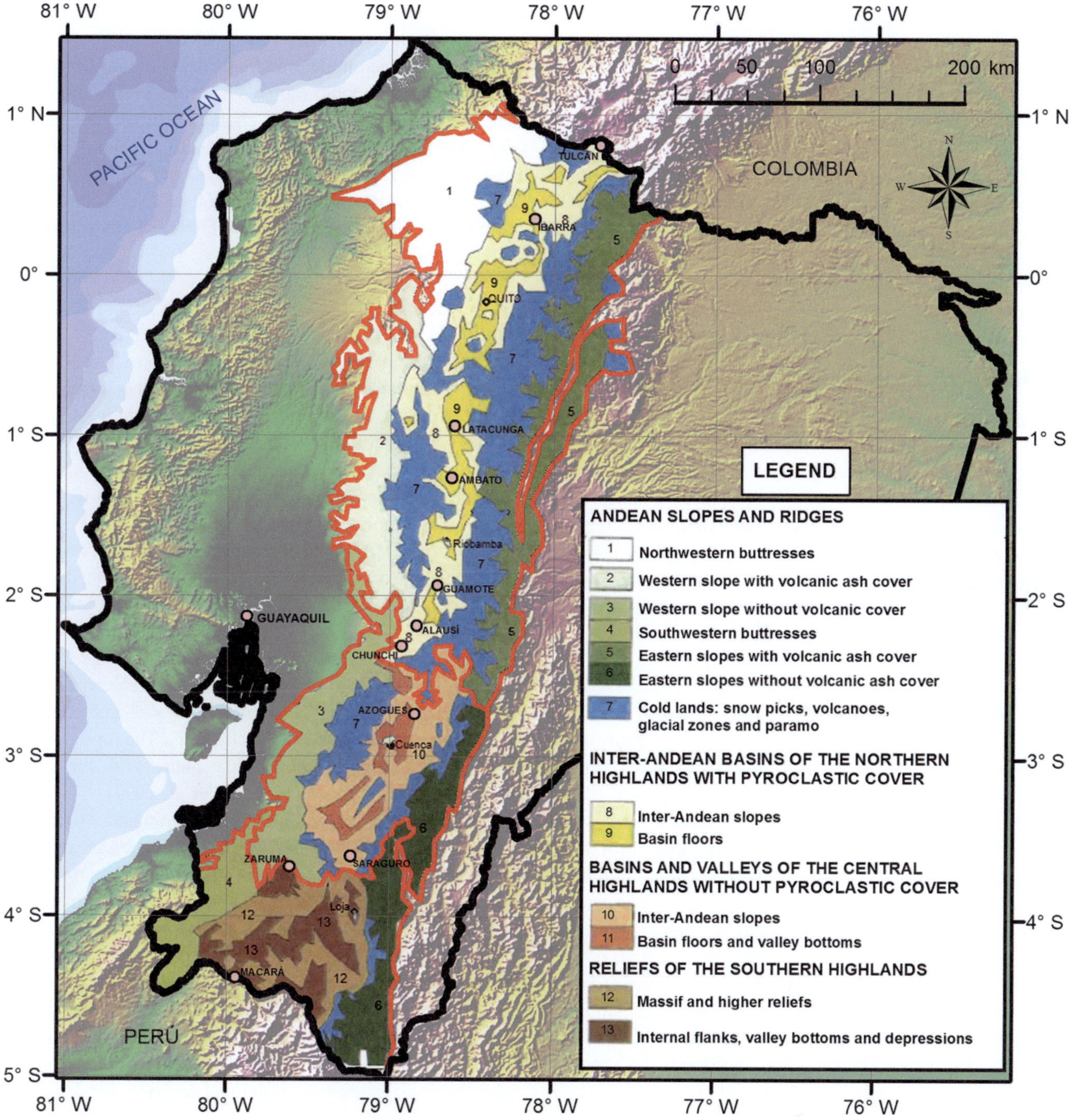

Fig. 3.2 Different landscapes from the Northern, Central, and Southern Highlands with respect to continental Ecuador (Winckell et al. 1997)

with vigorous modeling over highly dissected diverse ancient rocks. The climate is humid to wet with presence of heavy clouds, condition that favored the formation of perennial wooded zones. The soils are ferralitic with partial coverage of recent volcanic ash and rejuvenated by erosion; (c) Inter-Andean basins of the northern Sierra with volcano-sedimentary fillings, where the evident volcanic morphogenesis was promoted by the presence of large and active recent stratovolcanoes. The soils from this zone have developed a widespread pyroclastic coverage and are subject to intensive agriculture; (d) Inter-Andean basins of the central Sierra over ancient, volcanic and metamorphic base, without the presence of active volcanoes or recent pyroclastic covering. This zone is fragmented into two sets of parallel sedimentary basins with a relief distributed in a topo-climate-sequence where soils from upper part are ferralitic, often with high humus content, soils from the middle area are moderately evolved with mollic or vertic features, and soils from the lower part are poorly developed; (e) Notches and Inter-Andean fluvial valleys, with relative rich alluvial soils with diverse agricultural activity; (f) Reliefs of the Southern Sierra with fersialitic–fersallitic soils, located in a paleo-topo-climate soil sequence of large valleys and orthogonal depressions where there is a clear stratification from humid to arid climate (Winckell et al. 1997).

Land aptitude maps at scale 1:200,000 (PRONAREG-ORSTOM 1982) indicate that 23% of the land in the Highlands (1.4 million ha) is suitable for agricultural use. These are lands located on flat to moderately undulating areas of the intermountain valleys with favorable conditions for a wide range of temperate climate crops such as potatoes, barley, wheat, corn, quinoa, legumes, fruits, and vegetables. Twenty four percent (about 1.5 million ha) corresponds to very dissected areas with moderate to steep slopes, a condition that makes these lands marginal for agricultural use, but can be used for livestock production. Finally, 53% of the Highlands (3.4 million hectares) corresponds to areas with very steep slopes production (including external and internal slopes of the Andes) and serious climatic and morphopedologic constrains for agriculture and livestock. These areas can only be used for vegetation cover, forestation, and reforestation (Pacheco 2009).

3.2 Northern Highlands

The Northern Highlands or North Sierra stretches from the border with Colombia to latitude 2° 15′S at Palmira-Alausí (Fig. 3.3) The landscape of this unit is framed by two large parallel cordilleras (Eastern and Western) that form what Humboldt called The Volcanoes Avenue. These large massifs reach altitudes of 4000–4500 m a.s.l., and are separated by the Inter Andean depression which is about 20 km wide at altitudes ranging from 1600 to 3000 m a.s.l. Part of this unit are also the outside steep slopes that end up merging with the costal plane and the Amazonia. The volcanism from the Cretaceous period originated in the Eastern Cordillera, while the Western Cordillera was developed on Andean tectonic facies (Pacheco 2009; Winckell et al. 1997).

An intense volcanic process that developed over the original andesites and rhyolites generated large emissions of pyroclastic materials, lava flows and lahars that cover most of the Northern Highlands, area where the volcanoes are located in Ecuador, mostly without activity like the Cotopaxi colossus (5897 m a.s.l.) and Chimborazo (6310 m a.s.l.) (Fig. 3.4). Large volumes of volcanic ash deposited over the landscape, especially on the western cordillera and the coastal plain softened the landforms of these zones (Pacheco 2009).

Glaciation that occurred during the Quaternary, in combination with tectonic activity, Quaternary volcanism and deposition and erosion episodes, left a permanent print in the landscape and marked the current geomorphology of the Northern Highlands. Transverse volcanic edifices, called knots, were formed in the middle of the two cordilleras, as a consequence of tectonic faults. The knots divide the inside depression in the small valleys characteristic of the Northern Highlands (Pacheco 2009). The Northern Highland Sierra is divided into the following reliefs: (a) Cold summits; (b) Volcano flanks; (c) Outer flanks; (d) Inner flanks; (e) Bottom basin and inter-Andean alley reliefs; and (f) Fluvial environment.

3.2.1 Cold Summits

This Andean landscape, also known as Cold Lands of the Andes, is characterized by cold zones in the form of two meridional ribbons that crown the western and eastern cordilleras with protruding volcanic tops always covered with snow. This landscape begins at 0° latitude (San Antonio de Pichincha) at approximately 3600 m a.s.l., and in several places reaches heights with permanent snow, the highest is the Chimborazo peak which reaches 6310 m a.s.l. (Winckell et al. 1997). According to the Holdridge climate diagram, cold summits are among the subalpine, alpine, and nival altitudinal floors characteristic of the Boreal to subpolar and polar latitudinal regions (Cañadas 1983; Holdridge 1969).

When glaciations fully ended, recent glaciers appeared generating the current glacier morphology characterized by glacial valleys, moraines, and rocky cliffs. The topography is irregular with steep slopes and flat plains with the shapes softened by volcanic ash (Winckell et al. 1997).

The Western cordillera is of recent origin and was formed as a result of the crashing and crushing of the volcanic arcs that were blocked by the oceanic and continental crust when the Ecuadorian Coastal Plain ascended. The eastern cordillera was formed as a result of ancient Precambrian and/or Paleozoic activity, although recent studies indicate that the sedimentary and igneous rocks that cover this cordillera were formed in the Mesozoic. The cordilleras were covered with glaciers, in greater amounts in the eastern cordillera due to the higher volume of rainfall generated by the large Amazonian air masses (Derruau and de Noní 1982; Vallejo 2007; Winckell et al. 1997).

The climatic condition is the main feature that differentiates this zone from other Andean landscapes. Atmospheric pressure is low, humidity is high, rate of soil moisture is Udic and Perudic, environmental temperatures are low (<10 °C on average), soil temperature regime is generally Isofrígid, which means that the soil temperature at 50 cm is <10 °C. The average annual rainfall varies from 250 to 500 mm in the Cotopaxi and Tungurahua provinces, 500 to 1000 mm in the Imbabura and Chimborazo provinces, and 1000–2000 mm in the Pichincha, Carchi and Bolivar provinces (Cañadas 1983; Winckell et al. 1997).

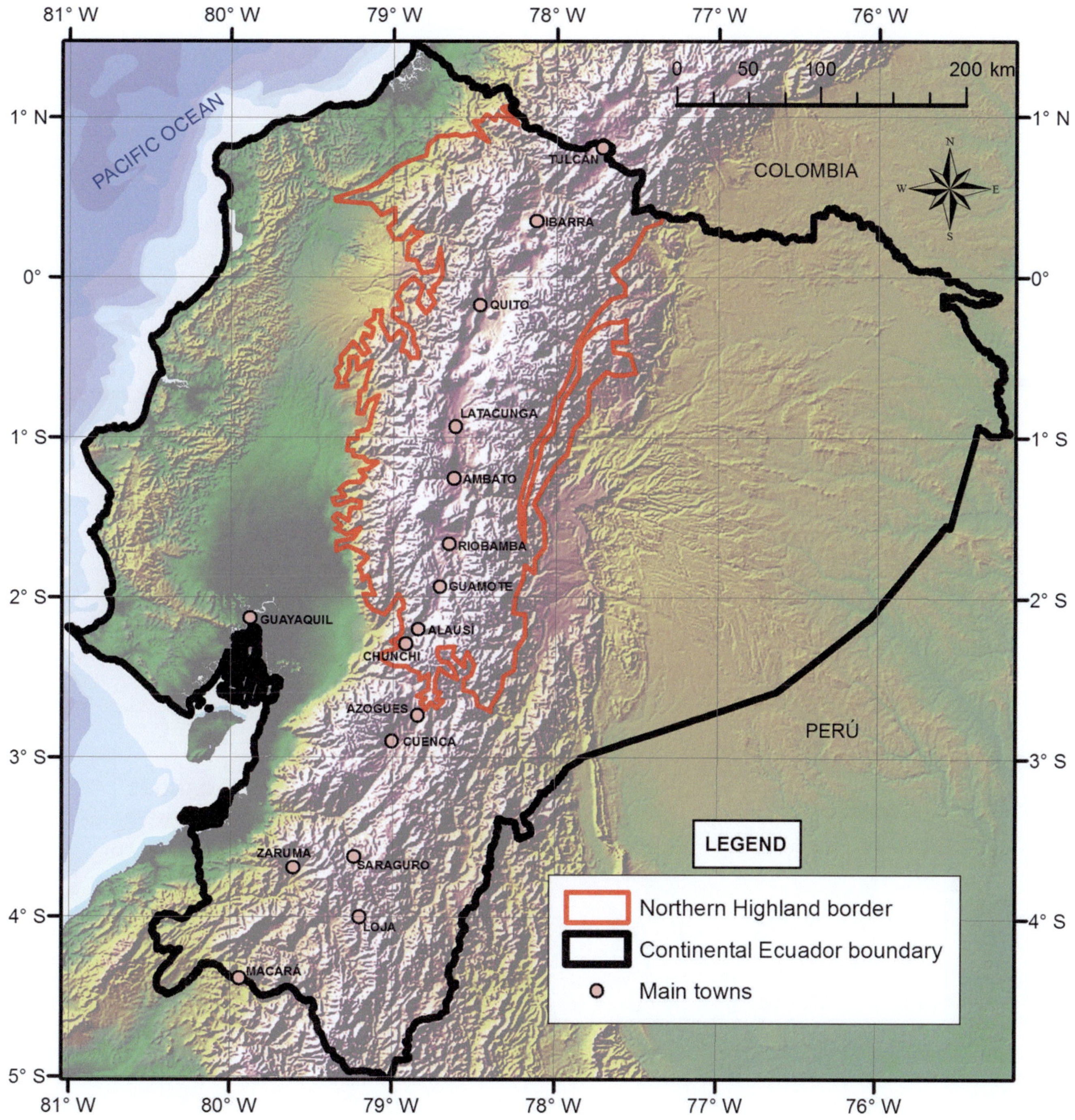

Fig. 3.3 Location of the Northern Highlands with respect to continental Ecuador (Pacheco 2009; Winckell et al. 1997)

The landscape of the cold summits is characterized by autonomous ecosystem called paramo that starts at the line of Andean forests and reaches the snow line. The paramo captures and controls freshwater that systematically drains through the vegetal cover and the colluvial alluvial soil providing the vital flow of water into the lower zones, a situation that sustains life and agricultural production (Beltrán et al. 2009; Poulenard et al. 2003). The water dividing line (Divortium Aquarum) is located on these summits and they give rise to the current large and small watersheds (Winckell et al. 1997).

The paramo vegetation is characterized by the presence of high altitude grasses (*Stipa ichu*), presence that gradually decreases as altitude increases changing to a vegetation of prostrate shrubs in the form of cushions or pads (*Plantago rigida*) along with remnants of *Polylepis* primary forest. Depending on the geographical zone (water availability, climate and volcano of origin), characteristic environmental

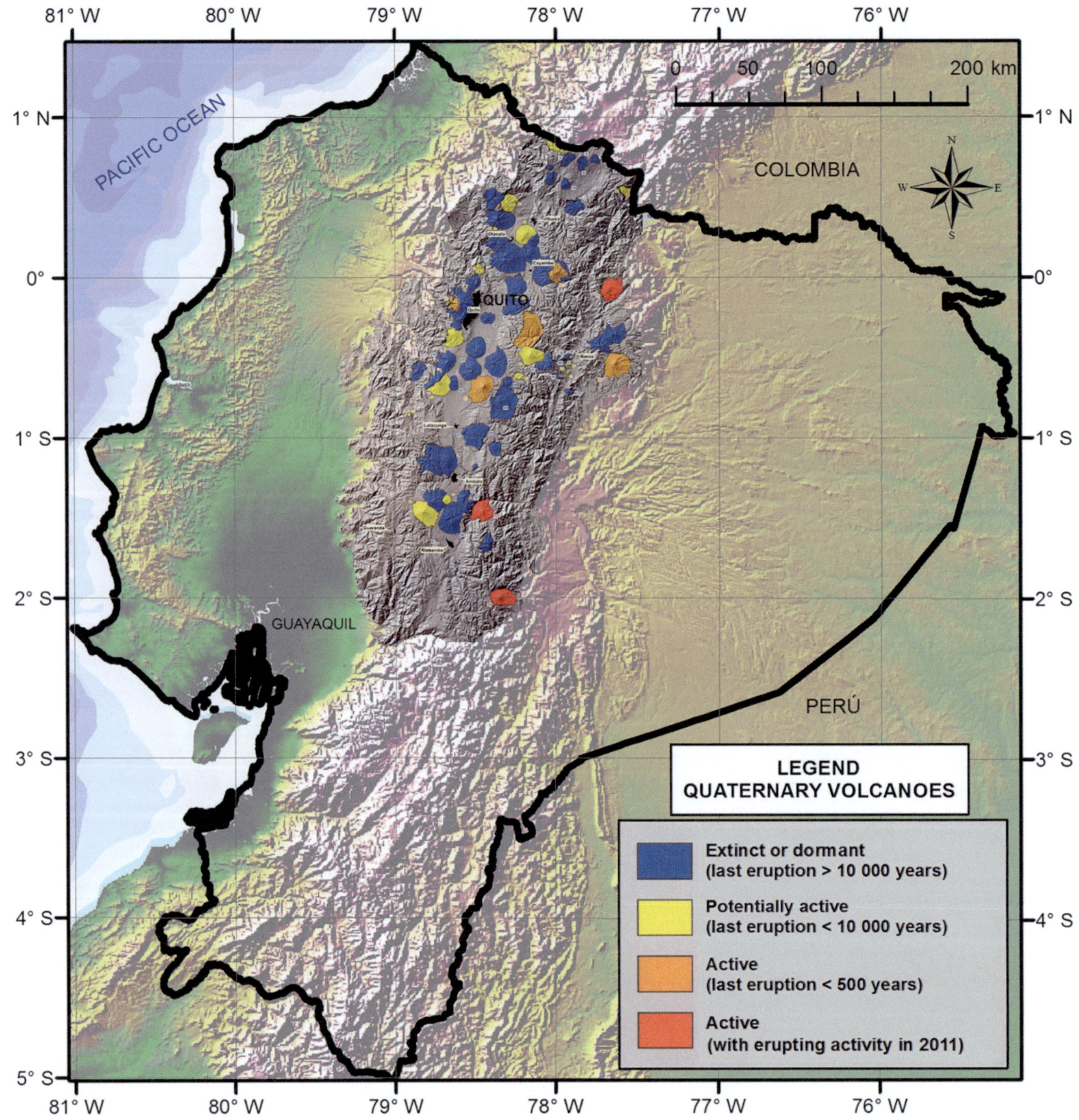

Fig. 3.4 Zone of influence of ancient and recent volcanic activity in Ecuador (IRD et al. 2012)

units can be differentiated with vegetal cover typical of each zone (Fig. 3.5) (Beltrán et al. 2009; de Bievre et al. 2011).

Due to its composition, the volcanic deposits of the cold summits weathered quickly and the large accumulation of humus, with the ability to complex the aluminum (Al), promoted the formation of stable humus-Al complexes. This strong Al complexation by humus limits the possibility of coprecipitation of Al and Si released from the weathering volcanic ash, which in turn limits the formation of allophane. For this reason, the soils of high altitude from the Northern Highlands are dominated by humus-Al complexes and have little allophane (Poulenard et al. 2003; Sollins 1991; Wada and Kakuto 1985; Zehetner et al. 2003). These soils are characterized for having dark colors, high carbon content,

Fig. 3.5 Volcanic modeling of cold peaks (left) and paramo grasses cover (*Stipa ichu*) (right) (IEE 2015c)

low base saturation, low bulk density, loamy, sandy loam or mainly silty loam texture, acid pH, and high amount of free Al^{3+} (MIDENA et al. 2013a).

Representative soils from the cold summits are classified mostly as Cryands (Soil Survey Staff 2006). Figure 3.6 and Table 3.1 describe the characteristics of a soil classified as Histic Cryaquands. This type of soil is located in valleys or glaciers surfaces with gentle slopes. The horizons have gley dark colors, histic features, and a high water table. The horizon sequence in the profile is as follows: Ao/Bg1/Bg2 (IEE 2015a).

Soils classified as Melanocryands and Fulvucryands are also found at the cold summits. Taxonomically, these soils are very young, very deep (in some areas the depth of the profile is >6 m), high in organic carbon, but with color differences possibly due to the dominance of humic or fulvic acids. The general sequence of the profile horizons is A1/A2/AC, but it has been documented that Melanocryands are located at higher altitude than the Fulvudands (MIDENA et al. 2013a). A modal profile of a soil classified as Melanocryands is presented in Fig. 3.7 (Table 3.2).

There are small areas of soils classified as Haplocryands (Fig. 3.8; Table 3.3) that are characterized by the presence of abundant volcanic glass (obsidian, pumice stone, and/or gravel), because they are located very close to the emission sites of magmatic material (crater or volcanic edifice).

Fig. 3.6 Profile of a soil classified as Histic Cryaquands (left) located in Pusin Pungo, Chambo, Chimborazo, covered with natural grass and pads (right) (IEE 2015c)

Table 3.1 Profile characteristics of a soil classified as Histic Cryaquands (IEE 2015c)

Horizons	Depth (cm)	Description
Ao	0–28	Brownish black (10YR 3/2) moist color, sandy loam texture, massive to subangular blocky structure, friable, pH 5.6, organic matter 18.8%, CEC 15 cmol kg^{-1}, base saturation 57.9%
Bg1	28–83	Dark gray (Gley 4/N) moist color, loam to sandy loam texture, angular blocky structure, friable, pH 5.5, organic matter 11.7%
Bg2	83–110+	Dark gray (GLEY 4/N) moist color, silty clay loam texture, massive to subangular blocky structure, friable

Fig. 3.7 Profile of a soil classified as Typic Melanocryands (left), located in Loma Mojanda, Pedro Moncayo, Pichincha, with its characteristic landscape (right) (IEE 2015c)

Table 3.2 Profile characteristics of a soil classified as Typic Melanocryands (IEE 2015c)

Horizons	Depth (cm)	Description
A1	0–50	Black (10YR 2/1) moist color, loam texture, granular to subangular blocky structure, slightly sticky and slightly plastic, strong reaction to NaF, pH 5.8, organic matter 12.9%, CEC 19 cmol kg^{-1}, base saturation 15.1%
A2	50–87	Black (10YR 2/1) moist color, loam texture, slightly sticky and slightly plastic, strong reaction to NaF
AC	87–110+	Brownish black (10YR 2/2) moist color, sandy loam texture, strong reaction to NaF

Paramo soils have a high potential to sequester CO_2, but are particularly important because they are the large water reservoir for the population located at the inter-Andean valleys (de Bievre et al. 2011). The conservation of these soils is a national priority.

3.2.2 Volcano Flanks

These Northern Highlands landscapes are related to the majority of the volcanoes present in the zone, and are defined around these structures, with moderate to severe slopes in a gradual transition to the Andean landscapes of lower areas. They are located between 3000 and 3500 m a.s.l., and vary in altitude depending on the volcano around which they are placed, especially at the south of the Northern Highlands. These areas have purely volcanic origin because the projections were deposited around the volcanic edifices and due to the slope were scattered around the landscape. Much of the ash was removed from the deposition site by the high water discharges that

Fig. 3.8 Profile of a soil classified as Vitric Haplocryands (left), located at Cerro Negro, Pedro Moncayo, Pichincha, with its characteristic landscape (right) (IEE 2015c)

Table 3.3 Profile characteristics of a soil profile classified as Vitric Haplocryands (IEE 2015c)

Horizons	Depth (cm)	Description
A1	0–50	Very dark brown color (10YR 2/2) when moist, silty loam texture, subangular block structure, slightly sticky and slightly plastic, strong reaction to NaF pH 5.1, 5.9% organic matter, CEC 8.0 cmol kg^{-1}, base saturation of 25.4%
A2	50–75	Black (10YR 2/1) moist color, loamy texture, subangular blocky structure, slightly sticky and slightly plastic, strong reaction to NaF
AC	75–90	Black (10YR 2/1) moist color, loamy sand texture with fragments of fine gravel, nonsticky and nonplastic, strong reaction to NaF
C	90–100+	Brownish black (10YR 3/2) moist color, massive structure with fine gravel fragments, strong reaction to NaF

accompanied to the volcanic eruptions. Through time, these episodes of volcanic material emission and dragging constructed profiles of alternating layers of different material (tuffs, lahars, flows of lava spills, etc.) with different sizes (ash, lapilli, pumice) rearranged in sectors of deep pyroclastic accumulation and sectors with rocky outcrops (Winckell et al. 1997). For these reasons, interfluvial units of flat to slightly wavy highpoints separated by marked rifts stretching hundred meters to several kilometers are observed. These areas have a hilly landscape, slopes of lava flows, and volcanic plateaus with heterogeneous profiles mostly with colluvial floors. The annual average temperature ranges from 11 to 16 °C, annual rainfall from 500 to 1,000 mm, and soil temperature regime is isomesic (MIDENA et al. 2013a).

Vegetation from the volcano flanks is diverse. There are sectors covered with primary vegetation and wildlife, indicating that they have not been altered. The typical paramo grasses are characteristic of several of these sites are, while in others there are dense scrubland, and arbustive vegetation. However, the lower parts of the flanks are occupied with pasture used for dairy activity or potato, faba bean or barley cultivation (MIDENA et al. 2013a). The advance of the agricultural frontier has been replacing the remaining paramo vegetation with pastures and crops, transformation that has significantly affected the conservation and buffer areas of the cold paramo (de Bievre et al. 2011).

The soils from the volcano flanks are black Andisols, immature, with poorly developed horizons, moderate to deep depth, and rich in organic matter. Most of these soils are classified as Melanudands and Hapludands (Soil Survey Staff 2006) (Figs. 3.9 and 3.10; Tables 3.4 and 3.5) (MIDENA et al. 2013b).

Fig. 3.9 Profile of a soil classified as Pachic Melanudands (left), located at La Parada, Tulcan, Carchi, with its characteristic landscape (right) (IEE 2015c)

Fig. 3.10 Profile of a soil profile classified as Typic Hapludands (left), located at Guangala, Cayambe, Pichincha, with its characteristic landscape (right) (IEE 2015c)

At the flanks also appear, in minor amount, soils classified as Mollisols, characterized by deep to moderately deep profile and with high organic matter content. Figure 3.11 and Table 3.6 describe a typical profile of a soil classified as Typic Argiudolls (Soil Survey Staff 2006).

3.2.3 External Flanks

These landscapes are located at the western and eastern outer sides of the great Andes massif. The western side is characterized by abrupt changes in slope as it descends to the

Table 3.4 Profile characteristics of the soil classified as Pachic Melanudands (IEE 2015c)

Horizons	Depth (cm)	Description
Ap	0–15	Brownish black (10YR 2/2) moist color, loamy texture, subangular blocky structure, slightly sticky and slightly plastic, moderate reaction to NaF
A	15–87	Black (10YR 2/1) moist color, sandy loam texture, granular to subangular blocky structure, slightly sticky and slightly plastic, strong reaction to NaF, pH 4.8, organic matter 15.1%, CEC 45 cmol kg^{-1}, base saturation 4%
AC	87–105	Brownish black (10YR 3/2) moist color, loam to sandy loam texture, massive to subangular blocky structure, slightly sticky and slightly plastic, moderate reaction to NaF, pH 5.9, organic matter 15.5%, CEC 58 cmol kg^{-1}, and base saturation 5%
2C	105–120+	Yellowish brown (10YR 5/8) moist color, sandy texture, without structure (massive), moderate reaction to NaF.

Table 3.5 Profile characteristics of a soil profile classified as Typic Hapludands (IEE 2015c)

Horizons	Depth (cm)	Description
A1	0–30	Brownish black (10YR 2/2) moist color, loamy texture, granular to subangular blocky structure, slightly sticky and slightly plastic, moderate reaction to NaF, pH 6.1, organic matter 5.8%
A2	30–110	Black (10YR 2/1) moist color, loamy texture, subangular blocky structure, sticky and plastic, little or no reaction to NaF

Fig. 3.11 Profile of a soil classified as Typic Argiudolls (left) located at Milan, Cayambe, Pichincha, with its characteristic landscape (right) (IEE 2015c)

coastal plain region, while the eastern side slopes gradually descend towards the Amazonia due to the transition break given by the Napo dome sub-Andean relief. The western slopes originated from the accumulation of volcanic material resulting from eruptions and dragging of a variety of sediments as a result of the intense tectonic activity endured by this area, while the eastern slopes developed on metamorphic formations which were covered by a blanket of pyroclastic projections of varying thickness, which were only modeled by erosion (Winckell et al. 1997). These areas are characterized by mountainous reliefs of diverse gradient, steep, abrupt slopes, dejection cones, colluviums, and undulating surfaces which are always separated by alluvial colluvium and deep ravines. These slopes form true barriers

Table 3.6 Profile characteristics of a soil classified as Typic Argiudolls (IEE 2015c)

Horizons	Depth (cm)	Description
Ap	0–33	Grayish yellow brown (10YR 5/2) moist color, loamy texture, subangular blocky structure, slightly sticky and slightly plastic, without reaction to NaF, pH 5.6, organic matter 5.9%, CEC 13.0 cmol kg^{-1}, base saturation of 60.1%
Bt1	33–65	Brownish black (10YR 3/1) moist color, clayey loam texture, subangular blocky structure, without reaction to NaF
Bt2	65–105+	Black (10YR 2/1) moist color, clayey texture, very sticky and very plastic, without reaction to NaF

Fig. 3.12 Profile of a soil classified as Thaptic Hapludands (left) located at Rio Sucio, San Miguel de los Bancos, Pichincha, with its characteristic landscape (right) (IEE 2015c)

walls which delineate the narrow and deep ravines by which the wide hydrographic system of the region circulates.

The two outer flanks are located in sectors varying from subhumid to very humid subtropical climates, with average temperatures ranging from 18 to 24 °C which are most of the time covered by clouds and constant fog (Cañadas 1983). The western slopes present annual precipitations ranging between 1500 and 3000 mm, but on the eastern slopes rainfall reaches as much as 5000 mm. The temperature and heavy rains on both slopes have promoted the development of an isohyperthermic soil temperature regime and an Udic soil moisture regime, Perudic in some sectors (Winckell et al. 1997).

The coverage of the external slopes is made of dense arboreal vegetation which includes the two major areas of ombrophilous forests of the country (northwestern and eastern flanks) with great floristic diversity and very short or nonexistent dry seasons. Strata with large trees pass to areas of shrubs as altitude increases, and finally, at the highest sites, the coverage changes to herbaceous plants (Huttel et al. 1999).

Main representative soils of these landscapes are Andisols, especially on the western slopes, where ashes from the Quaternary eruptions were distributed widely carried by alicious winds. These soils of dark brown color are desaturated, vitreous, and generally discontinuous in its vertical distribution (MIDENA et al. 2013a). A representative soil of this zone, classified as Thaptic Hapludands (Soil Survey Staff 2006) is presented in Fig. 3.12 and Table 3.7.

Soils classified as Inceptisols are also found in areas with high rainfall, high temperature, and steep slopes, conditions that promote continuous washing of exchangeable bases. The parent material shows signs of continuous weathering, evident in the oxidation of the profile (yellow to red colors). A representative soil of these zones, classified as Humic Dystrudepts (Soil Survey Staff 2006), is presented in Fig. 3.13 and Table 3.8.

Table 3.7 Characteristics of the soil profile classified as Thaptic Hapludands (IEE 2015c)

Horizons	Depth (cm)	Description
Ap	0–18	Dark brown (10YR 3/3) moist color, sandy loam texture, granular structure, strong reaction to NaF, pH 6.4, organic matter 4.5%, CEC 14 cmol kg^{-1}, base saturation of 15.4%
Bw1	18–45	Brown (10YR 4/4) moist color, sandy loam texture, subangular blocky structure, strong reaction to NaF, pH 6.6, organic matter 3.0%, CEC 14 cmol kg^{-1}, base saturation of 13.1%
Bw2	45–70	Dark brown (10YR 3/4) moist color, sandy loam texture, subangular blocky structure, strong reaction to NaF
2Ab	70–95	Dark Brown (10YR 3/3) moist color, sandy loam texture, subangular blocky structure, strong reaction to NaF, pH 6.9, organic matter 1.9%, CEC 14 cmol kg^{-1}, base saturation of 9.7%
2C	95–105+	Brown (10YR 4/4) moist color, loamy sand texture, massive structure, strong reaction to NaF, with the presence of silica crystals

Fig. 3.13 Profile of a soil classified as Humic Dystrudepts (left) located in Rio Verde, Lita, Imbabura, with its characteristic landscape (right) (IEE 2015c)

Table 3.8 Profile characteristics of a soil classified as Humic Dystrudepts (IEE 2015c)

Horizons	Depth (cm)	Description
Ap	0–20	Brownish black (10YR 3/2) moist color, clayey loam texture, subangular blocky structure, pH 6.0, organic matter 10.5%, CEC 21cmol kg^{-1}, base saturation 29.2%
Bw	20–35	Yellowish gray (2.5Y 4/1) moist color, clayey loam texture, subangular blocky structure, pH 5.9, organic matter 6.8%, CEC 20 cmol kg^{-1}, base saturation 26.9%
BC	35–75	Yellowish brown (10YR 5/6) moist color, clayey texture, massive to subangular blocky structure with common presence of coarse fragments (stones)
C	75–120+	Yellowish brown (10YR 5/8) moist color, clayey texture, massive structure with abundant coarse fragments (stones)

3.2.4 Internal Flanks

This great landscape is divided into two zones: (a) Upper Slopes, located at altitudes >3000 m a.s.l. limiting with the cold summits of the cordilleras; and (b) Lower Slopes, located around 2500–2600 m a.s.l. (Winckell et al. 1997).

Upper Slopes

The upper slopes have extensions of wavy crests shaped by the glaciers and covered by paramo vegetation. These are zones of frequent cloud cover, because the air masses coming from the west and east have lost much of their moisture by orographic precipitation in the external flanks and cold summits (Winckell et al. 1997).

The dominant soils in this zone are black Andisols, very unctuous, sometimes with the presence of a black clayey subsurface layer, known locally as chocoto (MIDENA et al. 2013a). A representative soil of this zone, classified as Typic Hapludands (Soil Survey Staff 2006), is shown in Fig. 3.14 and Table 3.9.

Lower Slopes

The lower slopes have very gentle gradients that are a gradual transition to the landscapes of the inter-Andean basins. The climate is different as the altitude changes, going from wet to dry as it descends from the high to the low areas, forming the system of microclimates characteristic of the areas inside the cordilleras which also controls the vegetative cover. This situation occurs because the flow of moist air coming from the coastal plain and the Amazonia must overcome the topographical obstacle that offers the great cordillera and on the road they unload much of their moisture on the external flanks and cold summits, and then descend into the inter-Andean zone as very dry air which picks up temperature in their descend to the bottom of the valleys (Cañadas 1983). For this reason, soil temperature regime of the Lower Slopes is isothermal and dominant soils are classified as Mollisols, characterized for having a dark color epipedon, rich in exchangeable bases and organic matter, all favorable conditions for root development (MIDENA et al. 2013a). The soils in these areas have layers of hardened material, called cangahua, at different depths, mainly formed by the sedimentation of volcanic tuffs. These soils are cultivated, among others crops, with corn, beans, wheat, potatoes, alfalfa, lupine, with the support of irrigation water. A representative soil of these zones, classified as Entic Durustolls (Soil Survey Staff 2006), is shown in Fig. 3.15 and Table 3.10.

These soils, generally deep, have suffered an accelerated erosion process that has left in many sites thin layers of soil over the hardened cangahua leaving little space for root development. For many areas, the cangahua has been fully exposed.

3.2.5 Bottom Basins and Inter-Andean Valleys

The two great volcanic chains of northern Ecuadorian Andes form in between a large alley around 350 km long that stretches from the border with Colombia to near the town of Alausí to the south. Internally, from east to west, several volcanoes stratum, called nodes, were formed, dividing this great alley in valleys and small basins. This entire unit is covered by soils derived from volcanic ash (Winckell et al. 1997).

The Bottom Basins and Inter-Andean Valleys, located at altitudes between 2000 and 3000 m a.s.l., have an annual average temperature ranging from 13 to 18 °C and an annual rainfall rate from 200 to 500 mm which delivers most of its discharge before and after the equinoxes having a marked dry season in the months of July, August, and September. In some sectors, the climate is arid (Cañadas 1983). These

Fig. 3.14 Profile of a soil classified as Typic Hapludands (left) located at Olmedo, Cayambe, Pichincha, with its characteristic landscape (right) (IEE 2015c)

Table 3.9 Profile characteristics of a soil classified as Typic Hapludands (IEE 2015c)

Horizons	Depth (cm)	Description
A1	0–50	Black (10YR 2/1) moist color, silty loam texture, subangular blocky structure, bulk density 0.77 g cm^{-3}, pH 5.5, organic matter content 8.5%, CEC 10 cmol kg^{-1}, base saturation 72.4%
A2	50–100+	Black (10YR 2/1) moist color, loamy texture, subangular blocky structure, pH 6.7, organic matter 5.8%

Fig. 3.15 Profile of a soil classified as Entic Durustolls (left) located at Titaicum, Chambo, Chimborazo, with its characteristic landscape (right) (IEE 2015c)

Table 3.10 Profile characteristics of a soil classified as Entic Durustolls (IEE 2015c)

Horizons	Depth (cm)	Description
Ap	0–30	Black (10YR 2/1) moist color, loamy texture, subangular blocky structure, pH 6.4, organic matter 4.4%, CEC 32 cmol kg^{-1}, base saturation 63.9%
CA	30–44	Dark yellowish Brown color (10YR 3/6) when moist, sandy loam texture, a massive to subangular blocks structure, with little presence of fine gravel
Cm	44–80+	Yellowish brown color (10YR 4/6) moist color, loamy sand texture without (massive) structure

conditions have promoted the development of an isothermic temperature regime and an Ustic soil moisture regime (MIDENA et al. 2013a).

The inter-Andean vegetation cover is varied and it is distributed in crop production systems by altitudinal floors. The production systems focused on potato and pasture for dairy production are located at the higher parts, while the lower and warmer areas crop systems include the cultivation of maize and beans, and in sectors with irrigation, vegetables, and fruits (Huttel et al. 1999). The areas with cut flower plantations (especially roses) grown under greenhouses have increased significantly in recent decades. These are highly profitable crops (PROECUADOR 2012).

Characteristic soils of these zones are classified as Mollisols, but have cangahua at a shallow depth. A representative soil of these zones, classified as Entic Durustolls (Soil Survey Staff 2006), is shown in Fig. 3.16 and Table 3.11.

In drier zones, soils are classified as Entisols and are characterized for having little or no evidence of pedogenic horizon development (MIDENA et al. 2013a). A representative soil of these zones, classified as Typic Ustipsamments is shown in Fig. 3.17 and Table 3.12 (Soil Survey Staff 2006).

Fig. 3.16 Profile of a soil classified as Entic Durustolls (left) located at Santa Rosa de Cusubamba, Cayambe, Pichincha, with its characteristic landscape (right) (IEE 2015c)

Table 3.11 Profile characteristics of a soil classified as Entic Durustolls (IEE 2015c)

Horizons	Depth (cm)	Description
A	0–30	Grayish yellow brown (10YR 4/2) moist color, sandy loam texture, subangular blocky structure, pH 7.2, organic matter 0.74%, CEC 8 cmol kg^{-1}, base saturation 91.9%
Cqm	30–55+	Brownish black (10YR 3/3) moist color, sandy loam texture, massive structure

Fig. 3.17 Profile of a soil classified as Typic Ustipsamments (left) located at Tocachi, Pedro Moncayo, Pichincha, with its characteristic landscape (right) (IEE 2015c)

Table 3.12 Profile characteristics of a soil profile classified as Typic Ustipsamments (IEE 2015c)

Horizons	Depth (cm)	Description
A	0–40	Dark brown (10YR 3/3) moist color, sandy loam texture, subangular blocky structure, pH 7.3, organic matter 0.3%, CEC 12 cmol kg^{-1}, base saturation 86.5%
2AC	40–55	Brown (10YR 4/4) moist color, sandy loam texture, massive to subangular block structure
2Cqm	55–80+	Yellowish brown (10YR 4/3) moist color, sandy loam texture, massive structure

3.2.6 Alluvial Environment

These are landscapes influenced directly by river activity. These are highly variable systems that change with time due to erosion and sedimentation, processes that also respond to climatic changes, modifications at the base level, Quaternary tectonics, and human activities. That is, these landscapes are the result of the continuous development of the Andean river network at different stages material undermining and deposition of the river beds. The vast majority are located at the lower parts of the Andean alley in dry climates (Winckell et al. 1997).

The agricultural use of the dry or the semiarid zones of the Alluvial Environments of the Northern Highlands is limited and only thorny shrubs species survive (penco and cactus). In places where there is access to irrigation, vegetable gardens and fruit trees are grown. In some places, there are eucalyptus forests (Huttel et al. 1999; Mejía 1997).

Soils of these dry zones, characterized by an alkaline pH and high base content, are classified mostly as Entisols. One example is the soil classified as Mollic Ustifluvents (Soil Survey Staff 2006) shown in Fig. 3.18 and Table 3.13.

3.3 Central Highlands

The Central Highlands extend from Palmira-Alausí at the north reaching the altitude of Zaruma-Saraguro at the south (Fig. 3.19). This unit has original features; it does not have recent volcanic edifices, it has developed extended monotonous plateau landscapes despite being a zone with a marked glacier footprint, and has an inherited relief from old and widespread volcanic spills. The Andes massif widens in this area reaching 40 km west of Cuenca, while the altitude progressively decreases towards the south, the highest point reaching only 3800 m a.s.l. west of Zaruma. The valleys of regular amplitude are distributed latter wise from 2500 m a.s.l. at Santa Isabel to 1600 m a.s.l. at Saraguro. These valleys open to the Coastal Plain through large incisions cut into the western flank by which the Huigra, Cañar, and

Fig. 3.18 Profile of a soil classified as Mollic Ustifluvents (left) located at Cajabamba, Colta, Chimborazo, on a high terrace on which carrots, broccoli and alfalfa are cultivated (right) (IEE 2015c)

Table 3.13 Profile characteristics of a soil classified as Mollic Ustifluvents (IEE 2015c)

Horizons	Depth (cm)	Description
Ap	0–20	Black (10YR 2/1) moist color, sandy loam texture, subangular blocky structure, pH 7.7, organic matter 3.1%, CEC 22 cmol kg^{-1}, base saturation 91.8%
C	20–60	Brownish black (10YR 2/2) moist color, sandy loam texture, massive structure, pH 7.9, organic matter 0.6%
2Ab	60–90	Brownish black (10YR 2/2) moist color, loamy texture, massive to subangular blocky structure, pH 8.1, organic matter 1.2%, CEC 18 cmol kg^{-1}, base saturation 96.2%
2C	90–105	Brownish black (10YR 2/2) moist color, sandy loam texture, massive structure

Jubones rivers are evacuated. These clefts facilitate the atmospheric circulation in the region (Winckell et al. 1997).

The Central Highlands do not have volcanoes or recent pyroclastic cover. The zone sits over an old base (volcanic and metamorphic) and the Inter-Andean area is fragmented into two sets of valleys and sedimentary basins distributed in parallel form (Fig. 3.20). The zone presents a pedologic cover distributed on a topo-climate sequence: soils at the higher areas are ferralitic–fersiallitic, often high in organic matter; soils at the middle areas are moderately evolved with mollic or vertic features, and soils at the lower areas are poorly developed (Winckell et al. 1997). Therefore, three landscapes or geosystems with different soil types have been identified in the Central Highlands: (a) Inter-Andean Basin Upper flanks; (b) Inter-Andean Basin Lower Flanks; and (c) Basin Bottom Reliefs.

3.3.1 Inter-Andean Upper Flanks

These landscapes are located mainly in the upper parts of large grooves of the Leon, Jubones, and Paute rivers. These are large steep slopes with a gradient >50%. The annual rainfall varies from 610 to 1285 mm, with dry periods lasting up to 6 months. The altitude of the lower limit varies from 1600 to 2800 m a.s.l., while the higher areas reach altitudes of 3200 m a.s.l. To the west, the rectilinear abrupts and escarpments are established over more coherent strata (lavas, agglomerates, and consolidated ignimbrites) of old volcanic formations, while to the east, the metamorphic landscapes are characterized by a dense and strong dissection in sharp crests separated by incisions with straight sides. Wetter areas, located at higher altitude, have soils with high organic matter and low base saturation (Fig. 3.21; Table 3.14), while at the lower and less humid zones (2200–2800 m a.s.l.) soils are fersiallitic, moderately deep and desaturated (Fig. 3.22; Table 3.15) (MIDENA et al. 2013c, g; Winckell et al. 1997).

This zone is occupied by large areas of grasslands in small properties and large extensive farms that reach the border with the paramo. In addition, there are smallholdings where barley, beans, potatoes, and corn are grown, and in the lower areas, corn and beans are grown (Winckell et al. 1997).

3.3.2 Inter-Andean Lower Flanks

These landscapes are a topographical continuity of the upper flanks, and gradually move down and diffuse into the bottoms of the valleys. These mountain flanks have softer slopes than the upper flanks. The climate of this zone is drier; with rainfall generally ranging from 475 to 750 mm with up to 10 months dry periods. Examples of these landscapes are those located at the bottom of the Oña basins (between 1800 and 2300 m a.s.l.) and Saraguro (between 2400 and 2800 m a.s.l.) as well as the gorges of the Paute basin. The main crops of these areas are corn, beans associated with corn, and wheat in the highest parts (Winckell et al. 1997).

The soils in these areas are clayey vertisols when the parent is made of metamorphic rocks (Fig. 3.23 and Table 3.16), and clay-stony when they are sedimentary rocks (MIDENA et al. 2013j; Winckell et al. 1997).

3.3.3 Basin Bottom Reliefs

The landscapes of this zone are located at lower altitudes and are less irregular. Given its geographical location, these are warmer and drier than the surrounding landscapes, however, there are variations in humidity in the stretch from 1000 to 3000 m a.s.l. ranging from very dry to wet. The wettest zones are at the Cuenca basin with 835 mm of annual rainfall and 2 dry months, while the valleys leading to the coastal plain, as the Cañar basin, are drier with only 500 mm of annual rainfall and 8 dry months. The basin bottoms are generally made of gentle slopes with a low relative gradient. An example of the soils of this zone is shown in Fig. 3.24 and Table 3.17. This soil, classified as Mollic Udifluvents

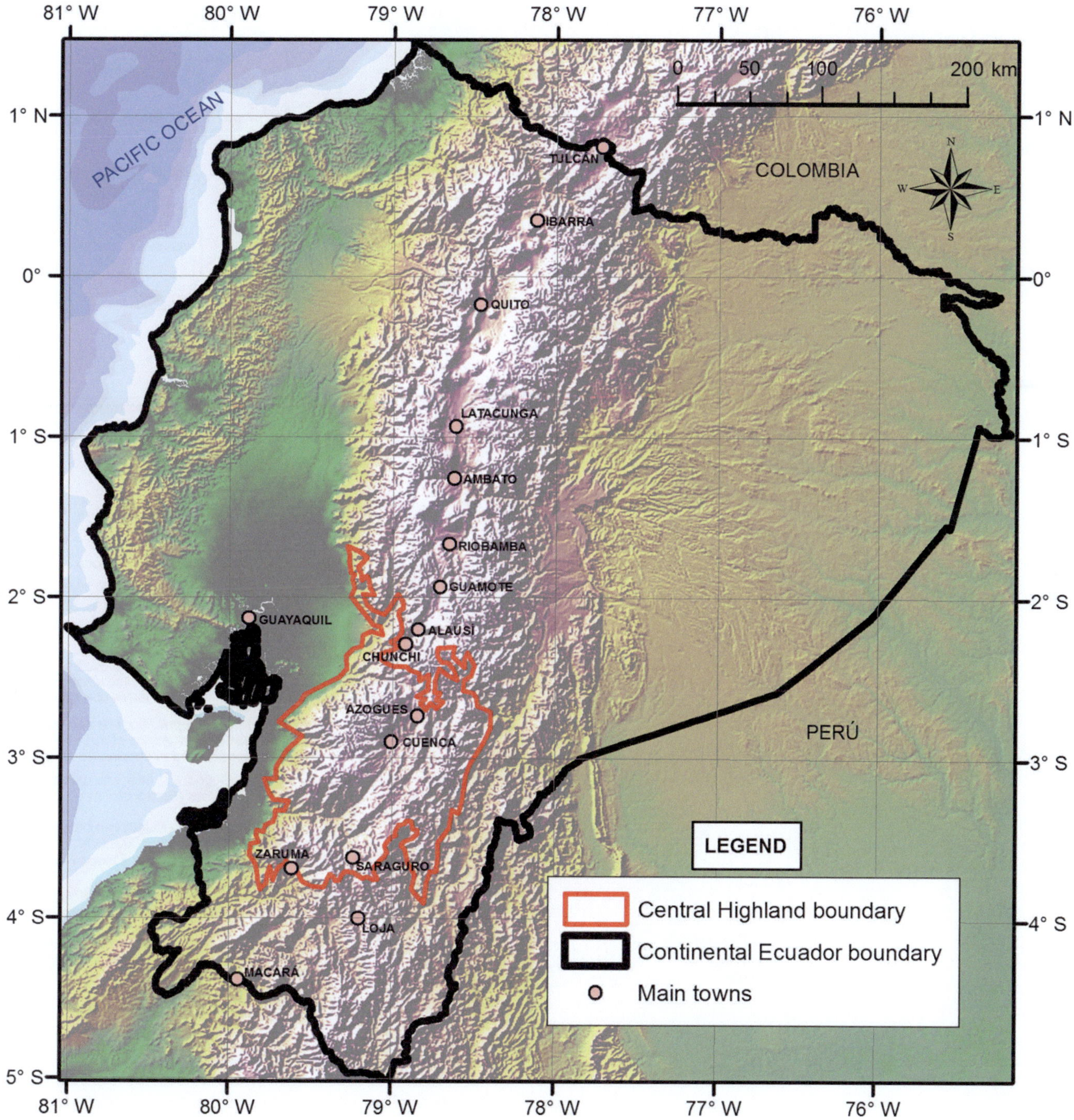

Fig. 3.19 Location of the Central Highlands with respect to continental Ecuador (Winckell et al. 1997)

(Soil Survey Staff 2006), is located at the bottom of the Paute River basin, where drier conditions promote a very high base saturation (>90%) (MIDENA et al. 2013j; Winckell et al. 1997). The dominant farming system is based on the typical corn-bean association and grasslands (Winckell et al. 1997).

3.4 Southern Highlands

Landscape delineation of the Southern Highlands is very particular. The first level of differentiation is climo-physiographic, because there is a high correlation between the type of soil with the climate gradient, condition that

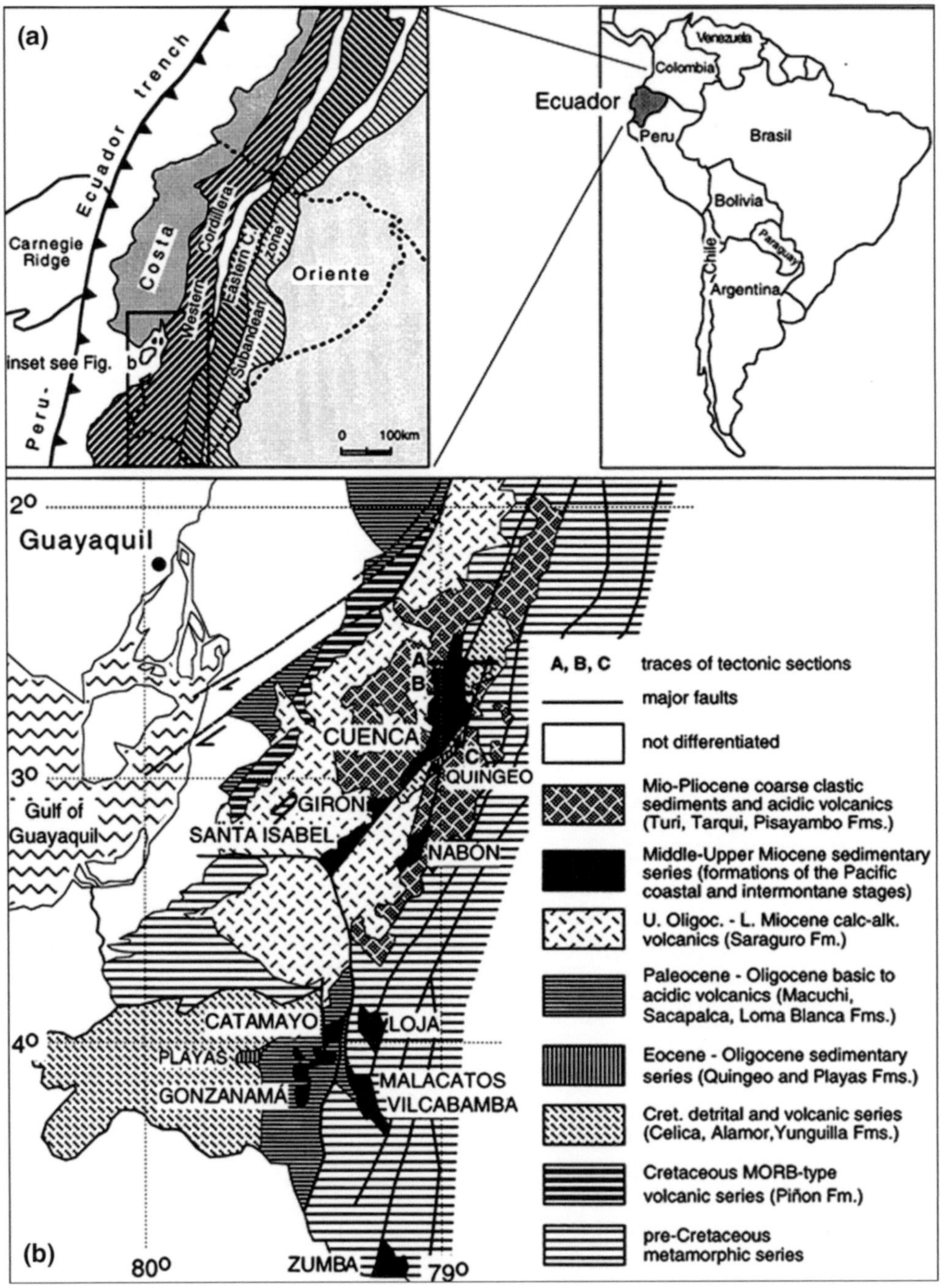

Fig. 3.20 **a** Morphotectonic subdivision of Ecuador; **b** Geological map of the Central and Southern Highlands highlighting the location of the main sedimentary basins (Steinmann et al. 1999)

arranges the landscapes into three morphopedologic groups: wet massifs, transition flanks, and drought prone lower zones. Second, it is also important to note the relationship between rock types and their weathering into soils to understand some of the apparent anomalies in the distribution of landscapes. This relationship divides the landscape into four main types: sandstones, metamorphic formations (volcanic and volcano-sedimentary), tertiary sediments, and glacis of colluvial deposits. The same way, several generations of alterites and consolidated soils, arranged in paleo-topo-climo sequences, cover the current landscapes in a sequence of fersalitic–ferralitic–fersalitic soils. In short, soils of the Southern Highlands have a climactic sequence overlying a paleo-sequence. This way, fersalitic, saturated, rich in 2:1 minerals with traces of 1:1 minerals soils were formed at the high zones, which gradually join with ferralitic desaturated paleo-soils. However, the lowlands are mainly dominated by Vertisols, often degraded the erosion,

Fig. 3.21 Profile of a soil classified as Vermic Udorthents (Soil Survey Staff 2006) (left) located on an ancient colluvium covered by silvopastoral systems (right) (IEE 2015c)

Table 3.14 Profile characteristics of a soil classified as Vermic Udorthents (IEE 2015c)

Horizons	Depth (cm)	Description
Ap	0–19	Brownish black (10YR 3/1) moist color, clayey texture, granular structure, pH 4.6, organic matter 14.3%, CEC 25 cmol kg^{-1}, base saturation 12.9%
AC	19–45	Yellowish brown (10YR 4/3) moist color, silty clay loam texture, massive structure, pH 4.9, organic matter 9.7%
Cr	45–70	Light yellow orange (10YR 8/4) moist color, unstructured

surrounded by fersalitic paleo-soils (Zebrowski and Sourdat 1997).

3.4.1 Relief Description

This unit is characterized by open reliefs, with large valleys and orthogonal depressions, leaving the Eastern Cordillera as the only line of continuous relief (Fig. 3.25). The landscape tops are high (>2800 m a.s.l.) and the highest point reaches 3791 m a.s.l. at the vicinity of the border with Peru. From this barrier, altitude decreases in west to northwest direction (Fig. 3.26). Shape of the landscape is always impressive with a prominent line of narrow string of glacier reliefs near watershed division. This great massif is only crossed by two gorges: the throat of Zamora River Valley at 1800 m a.s.l., and the southern gap at 2800 m a.s.l. (Winckell and Zebrowski 1997).

3.4.2 Geology and Climate

The Southern Highlands have not endured recent volcanic activity (post-Oligocene), typical of the Norther Highlands, situation that results in the absence of surface pyroclastic accumulation, therefore, the surface formations are derived exclusively from the weathering of the old substrate. The southern massifs have a lithology of metamorphic origin made of granitic, intrusive, volcanic, and volcano-sedimentary facies, complemented by the presence of some localized sedimentary deposits (Fig. 3.27) (Winckell et al. 1997).

The climate of the Southern Highlands is closely related to the altitude (Fig. 3.28). The tops of highest part are very wet, while the flanks are wet, but the amount of rainfall decreases as altitude decreases to a point that valleys and basins have a very dry to arid climate (MIDENA et al. 2013b, c, d, e; Winckell et al. 1997).

Fig. 3.22 Profile of a soil classified as Typic Dystrudepts (Soil Survey Staff 2006) (left), located on a high hilly relief with a slope <70% (right) (IEE 2015c)

Table 3.15 Profile characteristics of a soil classified as Typic Dystrudepts (IEE 2015c)

Horizons	Depth (cm)	Description
A	0–20	Brownish black (10YR 3/2) moist color, loamy texture, subangular blocky structure, laboratory data not available
Bw	20–35	Yellowish brown (10YR 4/3) moist color, sandy loam texture, subangular to angular blocky structure, pH 4.8, organic matter 4.1%, CEC 16 cmol kg^{-1}, base saturation 27.8%
BC	35–55	Brown (10YR 4/4) moist color, loamy texture, massive to subangular blocky structure
C	55–100	Yellowish brown (10YR 5/6) moist color, clayey loam texture, massive structure
R	100+	Parent material

Due to its particular characteristics, it is best to describe the soils of the Southern Highlands grouping them into three broad landscapes: (a) Massif higher sections; (b) Middle and lower flanks; and (c) Valleys and basin bottoms.

3.4.3 Massifs Higher Sections

The crests wet tops are not extensive and the influence of the Pacific or Amazonia air circulation promotes the presence of an evergreen forest, characteristic of the upper flanks of the Highlands. At the highest positions, the slopes are covered with scrubby vegetation, but moving down the landscape is covered by a complex association of tree, shrub or herbaceous vegetation associated with large eroded areas. The wetter sectors are dominated by grassland, while the warmer inter-Andean flanks are occupied by subsistence crops (corn-cassava association) and sugarcane for alcohol and panela. The western flank, completely exposed to the Pacific air circulation, is covered with cultivated grasses, tropical perennial crops (coffee, oranges, and sugarcane) along with subsistence crops like maize and cassava.

On the other hand, the intense paleo-morphological dynamism of these landscapes left indelible prints on their surface formations. These landscapes, inherited from a wetter period, are now completely recolonized by vegetation and are usually stabilized. These conditions (paleo-morphologicals) are clearly evident in the landscape by the presence of red land spots where the soils were developed over old ferralitic alterites or yellowish brown to

Fig. 3.23 Profile of a soil classified as Typic Haplusterts (Soil Survey Staff 2006) (left), located at Naunchún, Sigsig, Azuay on slopes <40% (right) (IEE 2015c)

Table 3.16 Profile characteristics of a soil classified as Typic Haplusterts (IEE 2015c)

Horizons	Depth (cm)	Description
Ap	0–17	Brownish black (10YR 3/2) moist color, clayey loam texture, subangular to angular blocky structure, pH 6.8, organic matter 1.1%, CEC 27 cmol kg^{-1}, base saturation 98.5%
A	17–35	Brownish black (10YR 3/2) moist color, clayey loam texture, subangular to angular blocky structure
Ab	35–75	Black (10YR 2/1) moist color, clayey texture, angular blocky structure
2C1	75–85	Grayish yellow brown (10YR 5/2) moist color, clayey texture, massive structure
2C2	85–95+	Grayish yellow brown (10YR 6/2) moist color, clayey texture, massive structure

whitish where they developed from current surface formations (MIDENA et al. 2013h; Winckell et al. 1997; Zebrowski and Sourdat 1997). The landscapes have relatively homogeneous climatic characteristics. The wettest areas, located near the eastern cordillera, have 1–2 dry months, the central sector 4 dry months and the western areas 6–8 dry months (MIDENA et al. 2013d, e; Winckell et al. 1997).

The geological heritage and the base lithology from this zone affected the distribution of the surface formations and soils, making common the soils developed from volcanic and sedimentary-volcano lithography, partially metamorphic, in addition to some soils developed from a granitic base (MIDENA et al. 2013h, i). Soils classified as Inceptisols and Ultisols (Soil Survey Staff 2006) predominate on the areas of volcanic and volcano-sedimentary rocks (IEE 2015a). An example of these Inceptisols is the soil classified as Dystric Eutrudepts (Soil Survey Staff 2006) (Fig. 3.29; Table 3.18), on which corn and peanuts are grown, products that stimulate the local economy, in alternate cycle with pasture in the dry season (IEE 2015b).

As indicated above, very old soils (paleo-soils) classified as Ultisoles (Soil Survey Staff 2006) have also been identified in this zone. They are characterized by having a Bt horizon, low base saturation, and high acidity, indicating that they were developed in a more humid climate than the prevalent today. A representative soil profile of these paleosols is shown in Fig. 3.30 and Table 3.19 (MIDENA et al. 2013j).

3.4.4 Middle and Lower Flanks

Due to its drier conditions, this intermediate section has facies which are covered with vegetation that ranges from relatively open shrubs to almost closed thorny plants. These areas are cultivated with corn associated with other

Fig. 3.24 Profile of a soil classified as Mollic Udifluvents (left), located at Pamar Común, Sigsig, Azuay with a slope <5% (right) (IEE 2015c)

Table 3.17 Profile characteristics of a soil profile classified as Mollic Udifluvents (IEE 2015c)

Horizons	Depth (cm)	Description
Ap	0–15	Brownish black (10YR 3/2) moist color, sandy clay loam texture, subangular blocky structure with evidence of coarse gravel, pH 7.2, organic matter content of 2.4%, CEC 38 cmol kg^{-1}, base saturation 96.9%
A	15–50	Brownish black (10YR 3/1) moist color, sandy clay loam texture, angular blocky structure and presence of medium size gravel, pH 7.2, organic matter 2.4%, CEC 38 cmol kg^{-1}, base saturation 96.9%
C	50–75	Brownish black (5YR 2/1) moist color, clayey texture, massive structure, abundant coarse fragments
2C	75–90	Brownish black (10YR 3/1) moist color, clayey texture, massive structure with abundant fine gravel
3C	90–97+	Grayish yellow brown (10YR 4/2) moist color, sandy clay loam texture, massive structure with presence of fine gravel

subsistence crops such as peanuts. Grasslands occupy larger areas at the higher less dry zones. In the lower drier areas, associations of corn, sugar cane, and cassava are common (IEE 2015b; MIDENA et al. 2013b; Winckell et al. 1997). In these landscapes (lower flanks) its evident the presence of highly eroded areas due to the poor vegetation cover as result of the dry to very dry climate and because they are located on slopes generally >40% (Winckell et al. 1997).

On the other hand, these landscapes can be subdivided in relative simple way, on one hand, parent rock plays a dominant role (landscapes on granite), while the climate gradient segregates the intermediate sections of dry conditions from the lower parts that are very dry (MIDENA et al. 2013b).

The predominant soils in this part of flanks are mainly Inceptisols, followed by Entisols. These soils have reddish, brown, or yellow hues, depending on the composition of the surface formations (MIDENA et al. 2013f; Zebrowski and Sourdat 1997). Figure 3.31 and Table 3.20 show a profile of a soil classified as Udic Haplustepts (Soil Survey Staff 2006), developed on old volcanic and metamorphic rocks. These Inceptisols are shallow, with a clayey loam texture, with the presence of medium size gravel throughout the profile (MIDENA et al. 2013f).

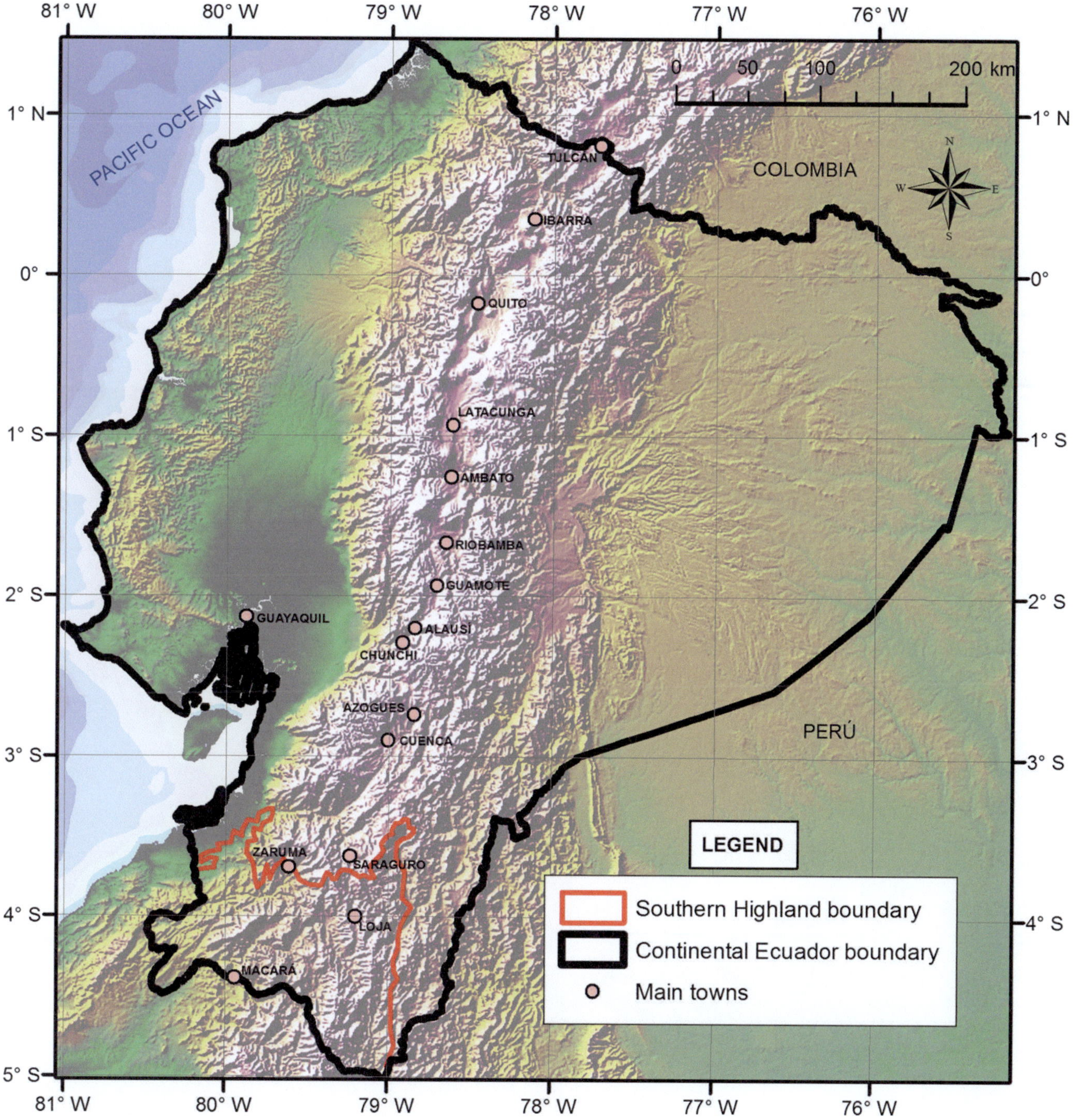

Fig. 3.25 Location of the Southern Highlands relative to continental Ecuador (Winckell et al. 1997)

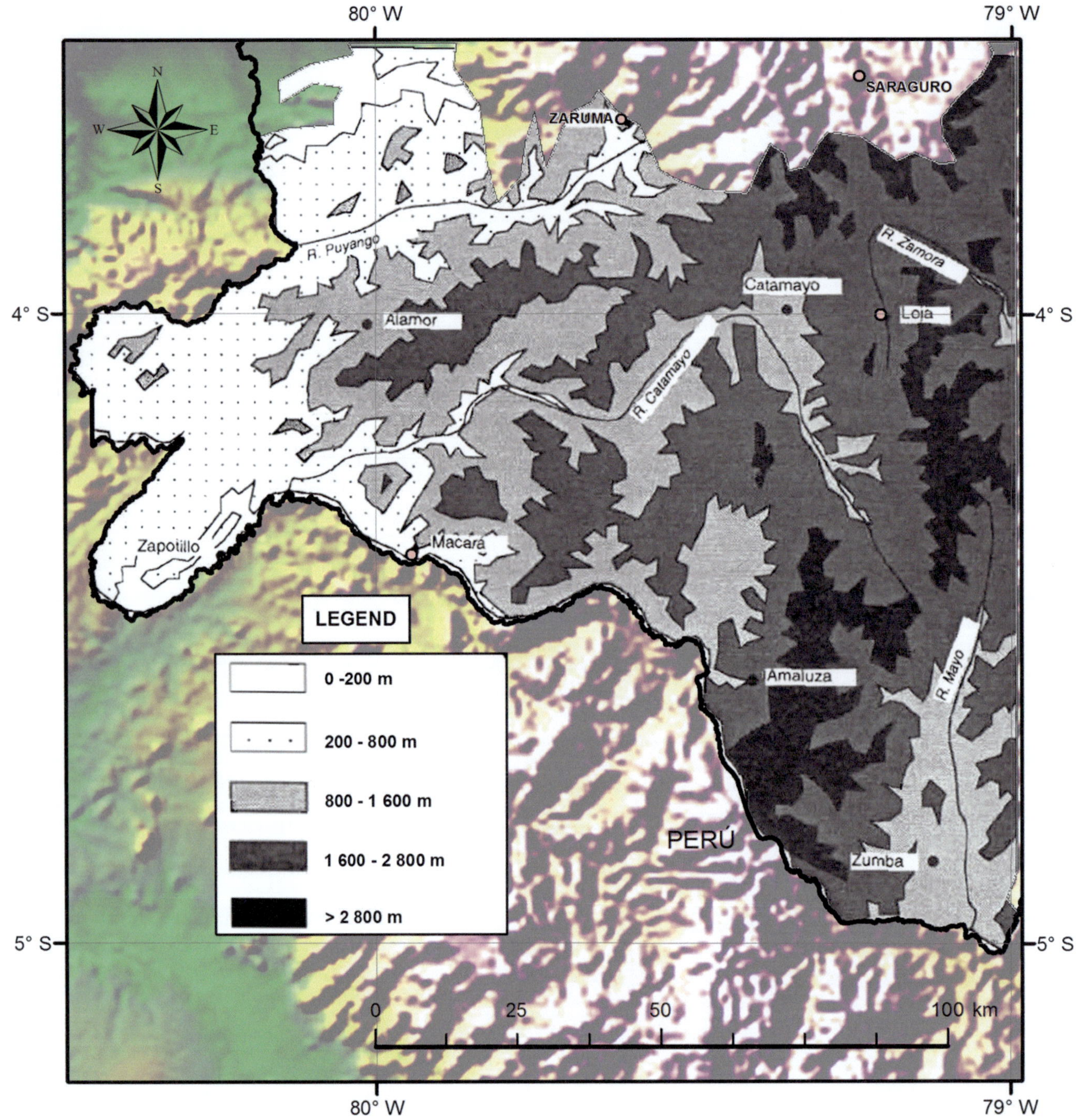

Fig. 3.26 Zones of altitude of the Southern Highlands (Winckell et al. 1997)

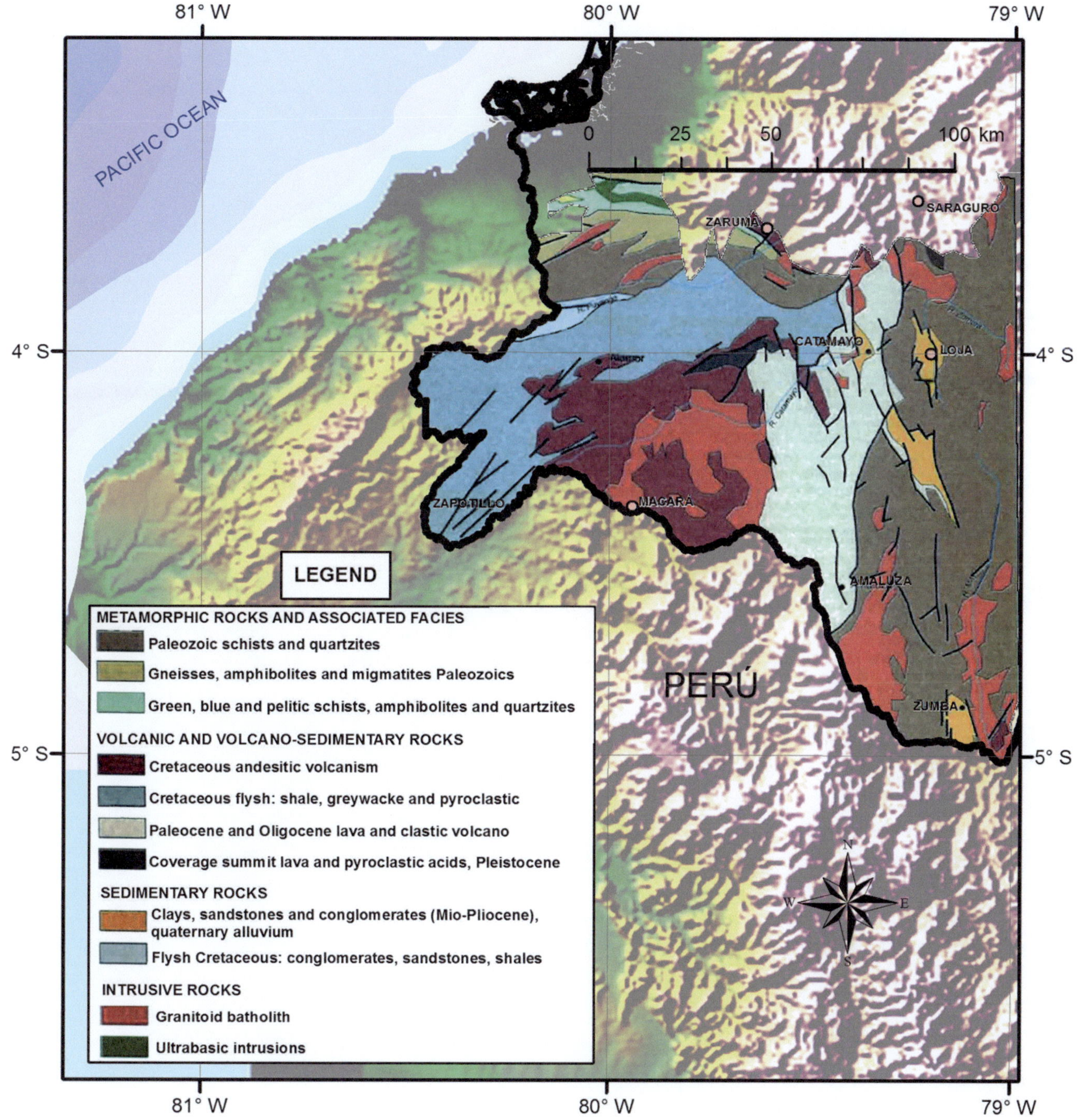

Fig. 3.27 Lithological formations of the Southern Sierra (Winckell et al. 1997)

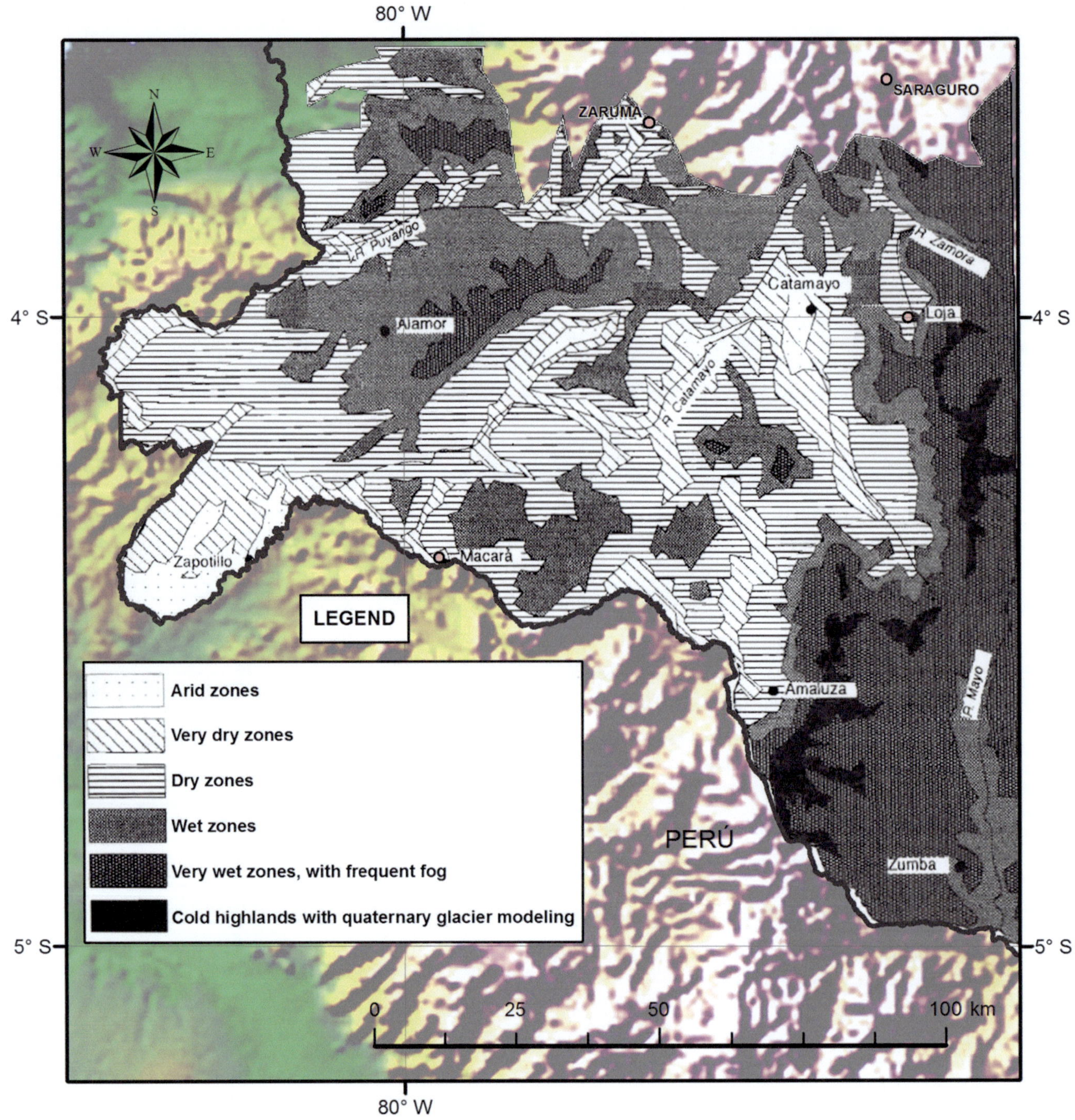

Fig. 3.28 Moisture zones of the Southern Highlands (Winckell et al. 1997)

Fig. 3.29 Profile of a soil classified as Dystric Eutrudepts (left) with brown color on the surface and yellowish red down below, located on a high hilly landscape at Santa Rosa, Olmedo, Loja (right) (IEE 2015c)

Table 3.18 Profile characteristics of a soil classified as Dystric Eutrudepts (IEE 2015c)

Horizons	Depth (cm)	Description
A	0–8	Brown (7.5YR 4/6) moist color, clayey texture, granular to subangular blocky structure, pH 6.3, organic matter 6.7%, CEC 23 cmol kg^{-1}, base saturation 73%
Bw	8–28	Reddish brown (5YR 4/6) moist color, clayey texture, subangular blocky structure, pH 6.1, organic matter 2.4%, CEC 18 cmol kg^{-1}, base saturation 72.4%
Cr1	28–65	Reddish brown (2.5YR 4/6) moist color, clayey texture, without structure (rocky)
Cr2	65–100+	Orange (7.5YR 6/8) moist color, clayey texture, without structure (rocky)

Fig. 3.30 Profile of a soil classified as Typic Paleudults (left) showing a Bt horizon with yellowish brown color. The soil is located in a mountainous relief covered with grasses at San Pablo, Paltas, Loja (IEE 2015c)

Table 3.19 Profile characteristics of a soil classified as Typic Paleudults (IEE 2015c)

Horizons	Depth (cm)	Description
A	0–12	Yellowish brown (10YR 4/3) moist color, clayey texture, subangular blocky structure, pH 5.5, organic matter 10.7%, CEC 23 cmol kg^{-1}, base saturation 28.2%
Bt	12–29	Dull yellowish brown (10YR 5/4) moist color, clayey texture, angular to subangular blocky structure with clay coating in the pores and few fragments of medium size gravel, pH 5.3, organic matter 5.6%, CEC 33 cmol kg^{-1}, base saturation 15.6%
BC	29–58	Bright brown (7.5YR 5/6) moist color, clayey texture, massive to subangular blocky structure, pH 5.4, organic matter 2.4%
C	58–70	Bright reddish brown (5YR 5/8) moist color, clayey texture, massive structure

Fig. 3.31 Profile of a soil classified as Udic Haplustepts (left), located at Tambo, Espíndola, Loja, cultivated with sugarcane (IEE 2015c)

Table 3.20 Profile characteristics of a soil classified as Udic Haplustepts (IEE 2015c)

Horizons	Depth (cm)	Description
Ap	0–24	Brownish black (10YR 3/2) moist color, clayey loam texture, subangular structure, pH 5.9, organic matter 2.9%, CEC 34 cmol kg^{-1}, base saturation 72.6%
Bw	24–42	Brown (10YR 4/4) moist color, clayey texture, subangular blocky structure, fragments of coarse gravel, pH 6.4, organic matter 1.6%, CEC 33 cmol kg^{-1}, base saturation 73.3%
C1	42–62	Dull yellowish brown (10YR 5/3) moist color, sandy clay loam texture, massive structure
C2	60–80	Greenish gray color (10 Gy 6/1) when moist, sandy clay loam texture, massive structure
C3	80–94+	Light olive gray (5G 7/1) moist color, sandy loam texture, massive structure

On the other hand, the Entisols of this zone are soils having continuous rock contact at 50 cm, condition which restricts the growth and development of different plant species, therefore, they are covered with dry herbaceous vegetation that fails to cover the soil and facilitates erosion processes. A representative soil, classified as Lithic Ustorthents (Soil Survey Staff 2006), is shown in Fig. 3.32 and Table 3.20 (MIDENA et al. 2013g).

Fig. 3.32 Profile of a soil classified as Lithic Ustorthents (left), located at an intermedian slope at Naranjillo, Loja (IEE 2015c)

Tabla 3.20 Profile characteristics of a soil classified as Lithic Ustorthents (IEE 2015c)

Horizons	Depth (cm)	Description
A	0–15	Brownish black (10YR 2/2) moist color, clayey loam texture, subangular blocky structure, pH 6.4, organic matter 6.9%, CEC 31 cmol kg^{-1}, base saturation 80.7%
AC	15–30	Brownish black (10YR 2/2) moist color, clayey texture, massive to subangular blocky structure, coarse fragments of gravel
Cr	30–45	Black (10YR 2/1) moist color, no developed structure
R	45+	Light gray (10YR 8/1) moist color, unstructured (rock)

3.4.5 Valleys and Basin Bottoms

The average annual temperature of the lower valleys is >22 °C, examples are Macará with 24.9 °C at 430 m a.s.l., and La Toma with 23.9 °C at 1240 m a.s.l., while the higher basins, such as Loja, have an average annual temperature is around 16 °C at 2140 m a.s.l. Overall, rainfall relates well to temperature, the more humid corresponds to the higher fresher parts, while the lowlands are dry or very dry. There is more variation in the valleys located in the middle flanks, there are dry and wet areas, for example, Rio Pindo with an annual average rainfall of 1465 mm (MIDENA et al. 2013c; Winckell et al. 1997).

The lower valleys, characterized by high river sediment depositions, are very dry to arid; conditions that only allow the development of adapted a spiny shrub vegetation cover. The superior margin of the lower basins is cultivated with corn in association with pastures, because climate conditions are less drastic. However, there is intense farming activity at the irrigated areas such as the Catamayo River Valley, Vilcabamba, Malacatos, and Catamayo, with production of sugar cane, bananas, citrus, coffee, cassava, and maize. The Catamayo valley, located on flat land of recent alluvial sediments, accommodates the large La Toma sugar mill and sugar fields irrigated with water from the rivers Guayabal and Catamayo (IEE 2015b; Winckell et al. 1997).

Dry climate and the erosion have promoted the presence of large areas with Entisols characterized by the outcrop of sandy, stony, and rocky formations. These are mineral sterile landscapes as the Playas River basin or upstream the Catamayo river valley, mainly desolated areas product of the dry to arid climate and the existence of stony sand or conglomerated detrital sedimentary base, which generates shallow, light soils (Winckell et al. 1997). On the other hand, at the Loja basin, the more humid climate has promoted the development of clayey soils classified as Vertisols (Soil Survey Staff 2006). An example of this type of soil, classified as Chromic Hapluderts, is presented in Fig. 3.33 and Table 3.21 (MIDENA et al. 2013g).

Fig. 3.33 Profile of a soil classified as Chromic Hapluderts (left) located at the site Tenería, Loja, Loja, on which grass are grown (right) (IEE 2015c)

Tabla 3.21 Profile characteristics of a soil classified as Chromic Hapluderts (IEE 2015c)

Horizons	Depth (cm)	Description
Apss	0–33	Grayish yellow brown (10YR 4/2) moist color, clayey loam texture, angular blocky structure, pH 5.2, organic matter 2.6%, CEC 22 cmol kg^{-1}, base saturation 62.5%
AC	33–60	Yellowish gray (2.5Y 5/4) moist color, clayey loam texture, massive structure, pH 5.2, organic matter 1.2%, CEC 26 cmol kg^{-1}, base saturation 63.4%
C	60–90	Brownish black (10YR 3/1) moist color, clayey loam texture, no structures, pH 4.9

References

Beltrán K, Salgado S, Cuesta F, León-Yánez S, Romoleroux K, Ortíz E, Cárdenas A, Velástegui A (2009) Distribución espacial, sistemas ecológicos y caracterización florística de los Páramos en el Ecuador. Quito, EcoCiencia, Proyecto Páramo Andino y Herbario QCA, Quito, Ecuador

Cañadas L (1983) El mapa bioclimático y ecológico del Ecuador. Editores Asociados, Quito, Ecuador

de Bievre B, Iñiguez V, Buytaert W (2011) Hidrología del páramo: Inportancia, propiedades y vulnerabilidad. In: Mena P, Castillo A, Flores S et al (eds) páramo. Abya Yala, Quito, Ecuador

Derruau M, de Noní J (1982) La geomorfología de los piroclastos en los andes del Ecuador. Geomorgología: documentos de investigación. CEG - IPGH - CEIG - ORSTOM, Quito, Ecuador

Holdridge L (1969) Life zone ecology. Tropical Science Center, San José, Costa Rica

Huttel C, Zebronski C, Gondard P, Bourliaud J (1999) Paisajes agrarios de Ecuador., vol V. Geografía Básica del Ecuador. Instituto Panamericano de Geografía e Historia, Quito, Ecuador

IEE (2015a) Geovisualizador Instituto Espacial Ecuatoriano. Mosaico de Geopedología 1:25 000. IEE. IEE. http://www.institutoespacial.gob.ec/geoportal/601-2/. Accessed 14 de Enero 2015

IEE (2015b) Geovisualizador Instituto Espacial Ecuatoriano. Mosaico de Uso de la tierra 1:25 000. IEE. http://www.institutoespacial.gob.ec/geoportal/601-2/. Accessed 14 de Enero 2015

IEE (2015c) Perfiles de suelos de Ecuador. IEE. http://www.institutoespacial.gob.ec/geoportal/perfiles-de-suelos/. Accessed 10 de Octubre 2015

IGM (2013) Atlas geográfico de la República del Ecuador. IGM - SENPLADES, Quito, Ecuador

IRD, IGM, EPN (2012) Volcanes cuaternarios del Ecuador continental. IGM, Quito, Ecuador

Mejía L (1997) Suelos del Ecuador: Reconocimento general en base a su capacidad-fertilidad y mapa general de clasificación por Capacidad-Fertilidad de los suelos del Ecuador: Una interpretación básica sobre las características de los suelos del Ecuador Fundación Forestal Juan Manuel Durini; Instituto de la Potasa y el Fósforo (INPOFOS) - Instituto Geográfico Militar (IGM) - Instituto panamericano de Geografía e Historia (IPGH) - Sociedad Ecuatoriana de la Ciencia del Suelo (SECS), Quito, Ecuador

MIDENA, SENPLADES, IEE, Gobierno Autónomo Decentralizado de Pichincha (2013a) Memoria técnica geopedológica del cantón Cayambe. Proyecto: Generación de geoinformación para la gestión del territorio a nivel nacional escala 1: 25 000. IEE, Quito, Ecuador

MIDENA, IEE, SENPLADES, MAGAP (2013b) Memoria técnica de clima e hidrología del cantón Espíndola. Proyecto: Generación de geoinformación para la gestión del territorio a nivel nacional escala 1: 25 000. IEE, Quito, Ecuador

MIDENA, IEE, SENPLADES, MAGAP (2013c) Memoria técnica de clima e hidrología del cantón Loja. Proyecto: Generación de geoinformación para la gestión del territorio a nivel nacional escala 1: 25 000. IEE, Quito, Ecuador

MIDENA, IEE, SENPLADES, MAGAP (2013d) Memoria técnica de clima e hidrología del cantón Olmedo. Proyecto: Generación de geoinformación para la gestión del territorio a nivel nacional escala 1: 25 000. IEE, Quito, Ecuador

MIDENA, IEE, SENPLADES, MAGAP (2013e) Memoria técnica de clima e hidrología del cantón Paltas. Proyecto: Generación de geoinformación para la gestión del territorio a nivel nacional escala 1: 25 000. IEE, Quito, Ecuador

MIDENA, IEE, SENPLADES, MAGAP (2013f) Memoria técnica de geopedología del cantón Espíndola. Proyecto: Generación de geoinformación para la gestión del territorio a nivel nacional escala 1: 25 000. IEE, Quito, Ecuador

MIDENA, IEE, SENPLADES, MAGAP (2013g) Memoria técnica de geopedología del cantón Loja. Proyecto: Generación de geoinformación para la gestión del territorio a nivel nacional escala 1: 25 000. IEE, Quito, Ecuador

MIDENA, IEE, SENPLADES, MAGAP (2013h) Memoria técnica de geopedología del cantón Olmedo. Proyecto: Generación de geoinformación para la gestión del territorio a nivel nacional escala 1: 25 000. IEE, Quito, Ecuador

MIDENA, IEE, SENPLADES, MAGAP (2013i) Memoria técnica de geopedología del cantón Paltas. Proyecto: Generación de geoinformación para la gestión del territorio a nivel nacional escala 1: 25 000. IEE, Quito, Ecuador

MIDENA, IEE, SENPLADES, MAGAP (2013j) Memoria técnica de geopedología del cantón SigSig. Proyecto: Generación de geoinformación para la gestión del territorio a nivel nacional escala 1: 25 000. IEE, Quito, Ecuador

Pacheco R (2009) El Ecuador: Recursos naturales agrícolas y del medio ambiente. Imprenta Colón, Quito

Poulenard J, Podwojewski P, Herbillon AJ (2003) Characteristics of non-allophanic Andisols with hydric properties from the Ecuadorian páramos. Geoderma 117(3):267–281

PROECUADOR (2012) Análisis sectorial de flores. PROECUADOR. http://www.proecuador.gob.ec/wp-content/uploads/downloads/2012/01/PROEC-AS2011-FLORES.pdf. Accessed 10 de diciembre 2014

PRONAREG-ORSTOM (1982) Inventario cartográfico de la Sierra ecuatoriana: mapas de aptitudes agrícolas, escala 1: 200 000, Quito, Ecuador. Ministerio de Agricultura y Ganadería (MAG), Programa Nacional de Regionalización Agraria (PRONAREG), Office de la Recherche Scientifique et Technique Outre Mer (ORSTOM), Quito, Ecuador

Soil Survey Staff (2006) Keys to soil taxonomy, 10th edn. USDA - NRCS, Washington

Sollins P (1991) Effects of soil microstructure on phosphorus sorption in soils of the humid tropics. In: Tielsen D, Hernandez D, Salcedo L (eds) Phosphorus cycles in terrestrial and aquatic ecosystems. Saskatchewan Institute of Pedology, Saskatoon, Canada

Steinmann M, Hungerbühler D, Seward D, Winkler W (1999) Neogene tectonic évolution and exhumation of the southern Ecuadorian Andes: a combined stratigraphy and fission-track approach. Tectonophysics 307:255–279

Vallejo C (2007) Evolution of the western cordillera in the Andes of Ecuador (late Cretaceous-Paleogene). Swiss Federal Institute of Technology Zürich, Zürich

Wada K, Kakuto Y (1985) Embryonic halloysites in Ecuadorian soils derived from volcani ash. Soil Sci Soc Am J 49:1309–1318

Winckell A, Zebrowski C (1997) Los paisajes costeros. In: Winckell A, Zebrowski C, Sourdat M (eds) Los paisajes naturales del Ecuador: Las regiones y paisajes del Ecuador. CEDIG - IPGH - IRD, Quito

Winckell A, Zebrowski C, Sourdat M (1997) Los paisajes naturales del Ecuador, vol IV Las regiones y paisajes del Ecuador. Centro Ecuatoriano de Investigación Geográfica (CEDIG) - IPGH (Sección Ecuador) - IRD (Francia) - IGM, Quito, Ecuador

Zebrowski C, Sourdat M (1997) Los factores de la pedogénesis y los suelos en Ecuador. In: Winckell A, Marocco R, Winter T et al (eds) Los paisajes naturales del Ecuador, vol Condiciones generales del medio natural. Centro Ecuatoriano de Investigación Geográfica (CEDIG) - IPGH (Sección Ecuador) - ORSTOM (Francia) - IGM, Quito, Ecuador

Zehetner F, Miller W, West L (2003) Pedogenesis of volcanic ash soils in Andean Ecuador. Soil Sci Soc Am J 67:1797–1809

Soils from the Amazonia

4

Darwin Sánchez, José Merlo, Renato Haro, Mayra Acosta, and Gustavo Bernal

Abstract

The Ecuadorian Amazonia is the part of the great Amazonia Basin, which borders with the Andes mountain range at the sector with more volcanic past. These landscapes extend from the Andes adjacent areas to the political border with Peru and Colombia to the east and northeast, respectively. This is a unique region which has been shaped by the arrangement of the geological structures and the climatic change that the region has suffered over geological times. The Amazonian reliefs are characterized by a gradual decrease in altitude from west to east resembling a set of stairs generating two characteristic landscapes: (a) Sub-Andean reliefs, which include corridors, mountain ranges and flanks; and (b) Peri-Andean Amazonia with Peri-Andean hills, and Piedmont and Fluvial Environments. In the Sub-Andean reliefs, the steep slopes range from 3000 to 600 m o.s.l., while the Peri-Andean reliefs are located at attitudes lower than 600 m o.s.l. Soils from the northern Sub-Andean reliefs are derived from volcanic ash, while at the south, without volcanic ash cover, old rocks have developed acid, reddish brown, light textured soils with high Al content. More than two-thirds of the Peri-Andean Amazonia has acid, heavy textured, highly leached, red, well-drained soils developed from tertiary sediments, while the northeast section developed over quaternary sediments covered by volcanic ash developed Andisols. The areas with fluvial influence have poorly drained, silty clay, greyey soils.

Keywords

Amazon basin • Parent material • Desaturated soils • Acidity and aluminum • Ferralitic and pseudoferralitic soils

4.1 Introduction

The Ecuadorian Amazonia is the part of the great Amazonia Basin, which borders with the Andes mountain range at the sector with more volcanic past. This is a unique region which has been shaped by the arrangement of the geological structures and the climatic change that the region has suffered over geological times (Custode and Sourdat 1986).

4.1.1 Geology, Geomorphology and Vegetation

The northwestern section of the Ecuadorian Amazonia region is characterized by the presence of high mountain

D. Sánchez (✉) · J. Merlo
Ministerio de Agricultura y Ganadería, Quito, Ecuador
e-mail: darwin.sanchez.rodriguez@gmail.com

J. Merlo
e-mail: jose.merlo@gmail.com

R. Haro · M. Acosta
Instituto Espacial Ecuatoriano, Quito, Ecuador
e-mail: renoharobull@gmail.com

M. Acosta
e-mail: mayritaacosta@gmail.com

G. Bernal
Asociación Nacional de Cultivadores de Palma Aceitera (ANCUPA), Quito, Ecuador
e-mail: gbernal@ancupa.com

J. Espinosa et al. (eds.), *The Soils of Ecuador*, World Soils Book Series,
https://doi.org/10.1007/978-3-319-25319-0_4

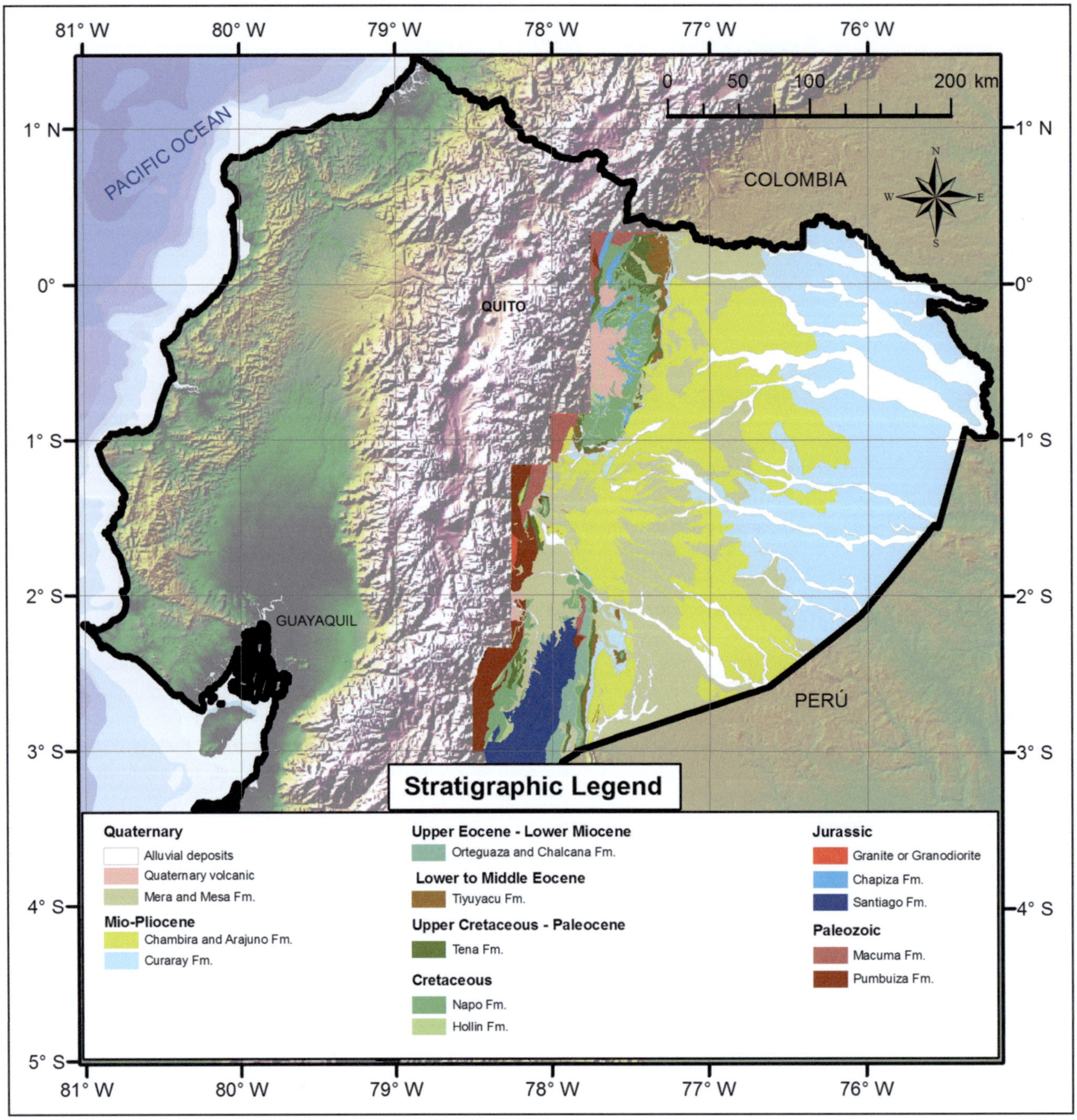

Fig. 4.1 Geological map of the Ecuadorian Amazon Basin: Provinces of Morona Santiago, Napo, Pastaza and Sucumbíos (IRD et al. 2001)

rocky reliefs with bassets of volcanic rocks, lava, sandy areas, volcanic tufa, and ash projections, while weathered schists and sandstone covered with volcanic ash are found towards the Peri-Andean slopes. Some structural reliefs are derived from Cretaceous sediments (Fig. 4.1). Primary forest, secondary forest, and areas where the natural vegetation has been replaced by grassland for livestock can be found in this zone (PRONAREG and ORSTOM 1983b).

In the center of the Ecuadorian Amazonia the presence of mostly granites covered by layers of ash is evident, however, moving toward the areas of greatest dissection, a block and boulder dedritic base of volcanic origin is present. Main limitations of the soils from this area are moisture excess and base leaching. Primary forest, secondary forest, small slash, and burn agricultural patches belonging to native tribes, and larger agricultural plots managed by settlers can be also found in this area (PRONAREG and ORSTOM 1983a).

In the Amazonia basin (altitudes <600 m o.s.l.), hills and plains can be differentiated. The hills were modeled by the dissection of the sedimentary banks. The presence of materials varying from totally weathered rocks to soft boulders altered into clay are evident in this area. The flat part is dominated by a slightly undulating relief developed over detrital material like boulders and sand. The predominant vegetation consists of forests and grasslands (Letamendi et al. 2007; PRONAREG and ORSTOM 1983a).

The soils in the southwestern part of the Amazonia region were mostly developed over metamorphic parent materials, with granitic intrusions and some sedimentary bassets, limited by tectonic fault lines. The Cutucú mountain range is characterized by an anticline model with asymmetric summits over Jurassic materials. The chevron formations correspond to Cretaceous and Tertiary materials. The predominant vegetation consists of forests, grasslands, and slash and burn patches of land managed by indigenous tribes (Letamendi et al. 2007; PRONAREG and ORSTOM 1983a).

The Ecuadorian southern Amazonia consists of a very rugged mountainous relief characterized by the presence of hills or summits over hard sandstone, which has left only small extensions for valleys. The lutites have formed little structured reliefs. The vegetation in this zone is mostly forests and grasslands (PRONAREG and ORSTOM 1983b).

4.1.2 Climate

The climate of Ecuadorian the Amazonia is homogeneous, condition that promotes the distinctive exuberant vegetation growth of this region. The climate is characterized by two rainfall gradients distributed as follows (Sourdat and Winckell 1997):

Rainfall Gradient from East to West

Data from the Amazonia weather stations demonstrate the direction in which this rainfall gradient develops. The Tiputini station, located at the eastern sector of the Ecuadorian Amazonia, registers an annual average rainfall of 2500 mm; the Limoncocha station, at the center of the region, 3245 mm; and the Tena station, at the foot of the eastern Andean flank, 3484 mm.

Rainfall Gradient from North to South

Sites with similar altitude register annual rainfall means that gradually decrease from north to south. At Cotundo, rainfall reaches values of 4590 mm, and at El Puyo 4410 mm; while continuing south, at Macas, a mean rainfall of 2550 mm is registered, and 2110 mm at Mendez. At the far south, at San Francisco and Cumbaratzá, a mean value of 1910 mm rainfall is recorded.

While the maximum amount of rainfalls during the period from March to July and the minimum from July to February, monthly variations are not significant. There are no dry months or water deficit throughout the region.

The annual average temperature of the zones located below 300 m o.s.l. is 25 °C. Zones between 1000 and 1100 m o.s.l., register an annual mean temperature of 20 °C. The variations of monthly means are <2 °C. The relative humidity varies from 85 to 90% (Espinoza Villar et al. 2009; Sourdat and Winckell 1997).

4.2 Soils from the Ecuadorian Amazonia

The Amazonian reliefs are characterized by a gradual decrease in altitude from west to east resembling a set of stairs, generating two characteristic landscapes (Sourdat and Winckell 1997): (a) Sub-Andean reliefs, which include corridors, mountain ranges and flanks; and (b) Peri-Andean Amazonia, with fluvial environments, Peri-Andean hills and Piedmont. These landscapes extend from the Andes adjacent areas to the political border with Peru and Colombia to the east and northeast, respectively (Fig. 4.2). In the Sub-Andean reliefs, the steep slopes range from 3000 to 600 m o.s.l., while the Peri-Andean reliefs, where humidity is permanent, the slopes are under 600 m o.s.l. (Custode and Sourdat 1986).

4.2.1 Sub-Andean Reliefs

These are the mountainous or sub-mountainous reliefs stepped from 2500 to 500 m o.s.l., place where the Andes eastern flank is located (Fig. 4.2). The landscapes correspond to the uprisings of anticlinal structure of corridors, depressions and adjacent foothills. Many of these landscapes are covered by volcanic ashes that have taken the shape of the already existing forms (Sourdat and Winckell 1997).

The geological formations of anticlinal uprisings, with marked alignment, are located between the rivers San Miguel and Anzú (Napo uprising) to the north, and between the Chinchipe and Pastaza rivers (Cutucú uprising) to the south (DGGM and NERC 1982). These anticline formations are composed mostly by secondary and tertiary sediments of marine or continental origin; hard rocks are present in the plateaus, slopes and spikes, while soft rocks are present in the tousled spikes or asymmetric high hills. Limestone (due to the lapiaz, avens structures), underground networks and sub-karstic forms are present in these formations. Much of

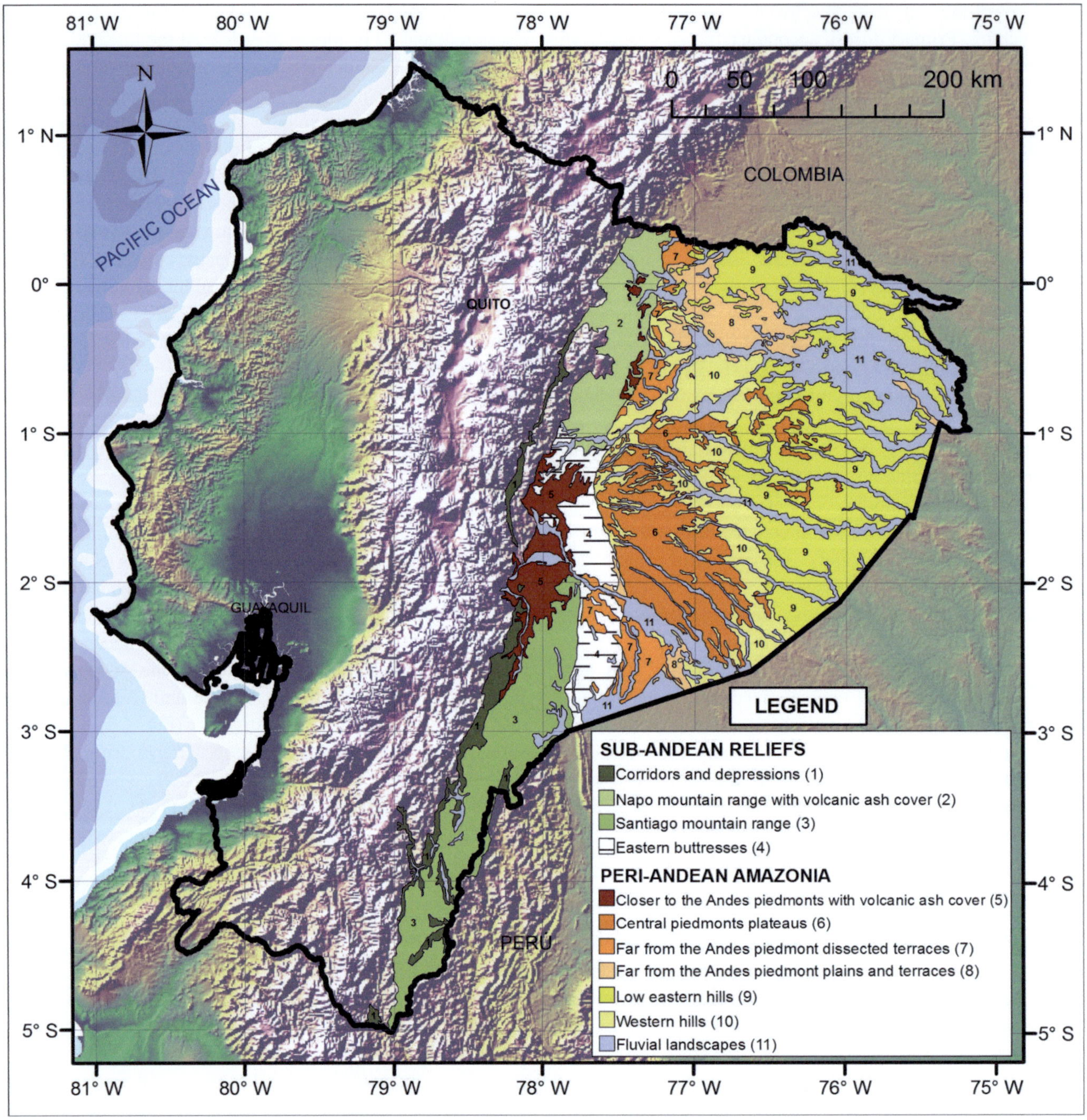

Fig. 4.2 Landscapes from the Amazonia with respect to the continental Ecuador (Sourdat and Winckell 1997)

these reliefs are covered with volcanic ash (Letamendi et al. 2007; Sourdat and Winckell 1997).

Corridors, Depressions, and Low Marginal Slopes

These units are divided by the presence of a volcanic ash coverage located to the north and center of the Sub-Andean reliefs (unit 1 of Fig. 4.2).

Reliefs with recent volcanic ash coverage: These reliefs are located at the Cosanga and Macas-Méndez corridors crossing from north to south the provinces of Napo, Tungurahua and Morona, in a stepped landscape with altitudes varying from 2000 to 500 m o.s.l., and a rainfall around 2500 mm. Soils classified as hydrated Andisols have developed over this area covered with volcanic ash. These soils are a mixture of ash and clayey colluviums over flysch and limestone (rocks of sedimentary origin) (Fig. 4.3) (Sourdat and Winckell 1997).

Fig. 4.3 Panoramic view of the Cosanga-Méndez Corridor characterized for having a cover of recent volcanic ashes (MIDENA et al. 2013a)

The modal profile of the corridors, depressions and marginal low slopes covered with volcanic ash (Baeza and El Chaco to the north and Macas and Sucúa to the center) corresponds to the soil classified as Hydric Hapludands (Soil Survey Staff 2006) (Fig. 4.4 and Table 4.1). These are moderately deep (72 cm) soils and have a strong reaction to sodium fluoride (NaF).

South of the corridor, soils are formed on deposits of boulder deposits of volcanic origin, and are classified as Oxic Dystrudepts (Soil Survey Staff 2006). These soils have brown color, low fertility status and high concentrations of Al (MIDENA et al. 2013a).

Reliefs without volcanic ash coverage: The Limón--Gualaquiza corridor is located between the Andes and the Cutucú and Cóndor mountain ranges, corresponding to the Limón Group structure (Fig. 4.5) which is characterized by diversity of shapes, landscapes and for having yellow soils. The Limón Group structure contains three different formations: (a) Formation with isolated peaks with ferralitic soils of medium organic matter content like those from the Zumba depression; (b) Peripheral metamorphic ridges, internal sedimentary deposits, slope lower parts, and colluvial slopes from the southern valleys with ferralitic soils, in some sites rejuvenated and stony, located over colluvial sediments; and (c) Highly dissected slopes with neutral and saturated ferralitic and pseudo-ferralitic soils rejuvenated by colluviation, developed on flysch and calcareous rocks (PRONAREG and ORSTOM 1983a).

Representative soils from this zone are classified as Lithic Dystrudepts and Oxic Dystrudepts (Soil Survey Staff 2006). In general, these soils have the following sequence of horizons: Ap/Bw/C. The surface horizons are black, while deeper horizons are yellow to reddish. Texture of the surface horizon is clayey loam and silty clay in the underlying horizon (Bw). Horizons A and B exhibit sub-angular to angular blocky structure. Chemically, soils have acidic pH, and base saturation <50% (MIDENA et al. 2013a). The Lithic Dystrudepts soils are shallower, since they have a lithic contact at 50 cm (Fig. 4.6 and Table 4.2).

Sub-Andean Mountain Ranges Between the Andes and the Amazonian plain a third mountain range, named Sub-Andean region, arises. The most evident reliefs are the dome of the Napo to the north and the Sierra of Cutucú to the south. These structures are separated by a depression located between the valleys from the upper Napo and the upper Pastaza. The Cóndor cordillera continues south and enters Peruvian territory (units 2 and 3 in Fig. 4.2) (Bristow and Hoffstetter 1977).

Napo mountain range with volcanic ash cover: These landscapes belong to the lifting of the Napo structure formed by Cretaceous sediments, mostly marine (unit 2 of Fig. 4.2). This unit is covered by volcanic ash. Components of these landscapes are sandy and clayey hilly reliefs, inclined or horizontal surfaces, edges and plateau flanks, hill flanks or chevrons, and surfaces and abrupt slopes from dejection and dispersion cones. There are also easy undulations over limestone, with sub-karstic modeling as lapiaz, sinkholes, and caves (Fig. 4.7) (Sourdat and Winckell 1997).

Fig. 4.4 Profile of a soil classified as Hydric Hapludands located at the corridors, depressions and low marginal slopes covered with volcanic ash (IEE 2015)

In these landscapes, the ashes from the Sumaco and Reventador volcanoes have covered all the preexisting forms with continuous layers, several meters thick. The pedogenic conditions of these ashes are similar for all landscapes, thus, average annual temperatures range from 17 to 24 °C depending on the altitude; rainfall varies from 4000 to 7000 mm; and the cloudiness is particularly strong. These climatic conditions explain the complete weathering of the ash mantle, regardless their age or depth (Sourdat and Winckell 1997).

The same authors indicate that this area is covered by dense tropical forest, which, exceptionally, adapts to the altitude or to the shallow soils developed on rough landforms. In the wetlands, this vegetation gives way to palm trees formations. The vegetation maintains appreciable litter layers over the ground which promotes the development of dense organic matter rich surface horizons, especially at high altitude. There is a partial conversion of forest to crops and pastures in this area (Fig. 4.8).

The soils developed on these landscapes are classified as hydrated Andisols, characterized by the predominance of amorphous clays, which, together with humus, form organic-mineral stable complexes, with a base saturation <50%, water retention capacity at pF3 > 100%, and very low bulk density (0.2–0.4 g cm^{-3}) (Custode and Sourdat 1986; Sourdat and Winckell 1997).

The horizon sequence of the modal profile for this zone is A/Bw/C, and belongs to a soil classified as Typic Hydrudands (Soil Survey Staff 2006) (Fig. 4.9 and Table 4.3). Reaction to NaF is very strong throughout the profile, except for the white horizon (rich in halloysite) (Zebrowski and Sourdat 1997). The organic matter content is also high (>10%) throughout the profile (Custode and Sourdat 1986).

Santiago mountain range without ash cover: The Santiago mountain range is bulkier and rugger than the Napo. In this structure, the Cutucú and Cóndor mountain ranges were formed and both of them are part of a third lower cordillera that runs parallel to the two main Andes cordilleras (unit 3 Fig. 4.2). Very rugged reliefs of flanks and chevron fronts, and dissected reliefs such as rectilinear and heterogeneous

Table 4.1 Description of the profile characteristics of the soil classified as Hydric Hapludands (IEE 2015)

Horizons	Depth (cm)	Description
Ap	0–20	Black (7.5YR 2.5/1) moist color, sandy loamy texture, sub-angular blocky structure, bulk density 0.31 g cm^{-3}, strong reaction to NaF, pH 5.5, CEC 20 cmol kg^{-1}, base saturation 21%, and organic matter 17.3%
A	20–50	Dark brown (7.5YR 3/2) moist color, sandy loam texture, sub-angular blocky structure, strong reaction to NaF, pH 5.7, CEC 15 cmol kg^{-1}, base saturation 17.9%, and organic matter 11.8%
Bw	50–72	Dark brown (7.5Y 3/4) moist color, sandy loam texture, sub-angular blocky structure, strong reaction to NaF, pH 5.4, and organic matter 6.5%
2C	72–100	Grayish brown (2.5Y 5/2) moist color, silty clay texture, massive structure

Fig. 4.5 Panoramic view of the Limón-Gualaquiza corridor without a volcanic ash cover

Fig. 4.6 Profile of soils classified as Lithic Dystrudepts (left) and Oxic Dystrudepts (right

Table 4.2 Description of the characteristics of the soil profile classified as Lithic Dystrudepts

Horizons	Depth (cm)	Description
Ap	0–15	Black (10YR 2/1) moist color, clayey loam texture, sub-angular blocky structure, pH 5.9, CEC 11.1 cmol kg^{-1}, base saturation 15.3% and organic matter 6.5%
Bw	15–40	Olive brown (2.5YR 4/6) moist color, silty clay texture, sub-angular and angular blocky structure, pH 4.9, CEC 18 cmol kg^{-1}, base saturation 16.4% and organic matter 3.8%
R	40+	Rock

Fig. 4.7 Landscapes of the Cordillera of Napo with a cover of volcanic ash (photo Darwin Sánchez)

slopes with strong dissection (massive with remains of structural forms) predominate in this unit (Fig. 4.10), and both are seated on limestone and sandstone. There are also residual monoclinal surfaces more or less dissected as the limestone inclined surfaces (PRONAREG and ORSTOM 1983a; Sourdat and Winckell 1997).

The Cóndor mountain range is an extension of the eastern main Andean cordillera, and extends about 150 km from north to south, forming part of the international border between Ecuador and Peru (Fig. 4.11). The highest altitude reaches around 2500 m o.s.l. The dominant landscape consists of rugged reliefs, mainly over granite and structural sandstones, such as ledges and plateau flanks. There are also surfaces of dissected cones, and hilly reliefs (Sourdat and Winckell 1997). These reliefs are lithologically associated to rocks from the Cretaceous period (Bristow and Hoffstetter 1977).

Both cordilleras are not covered by ash; with the exception of some localized areas in the surrounding slopes of the Sangay volcano which have an ash cover that it is assumed comes from the activity of this volcano. The altitudes of both mountain ranges vary from 2500 and 500 m o.s.l., therefore, mean annual temperatures are between 13 and 21 °C and annual rainfall is around 2000 mm (Sourdat and Winckell 1997).

These landscapes are covered by bushes, rainforests or tropical moist forests, which, unfortunately, tend to be reduced by slash and burn practices, annual crop planting (like maize) and the expansion of the pasture areas (Fig. 4.12). In recent years, the expansion of the agricultural areas, and mainly the areas dedicated to grazing has been one of the main agents accounting for the deterioration of the Ecuadorian Amazonian forests (Brondízio et al. 2009; Mena et al. 2011; PRONAREG and ORSTOM 1983b).

The edaphologic cover of this unit comprises the following soil types: (a) Poorly developed soils, (b) ferralitic and pseudo-ferralitic soils, and retouched podzols.

- **Poorly developed soils**: These are soils in which the erosion processes are higher than the pedogenic processes, therefore, the profiles are shallow, often <10 cm, generally with the presence of rock at a 50 cm depth, forming the following profile sequences: A/C/R or A/R. The surface horizon color is dark yellowish brown to yellowish brown, with clayey loam and clayey texture (Custode and Sourdat 1986). These soils are classified as Typic Udorthents and Lithic Udorthents (Soil Survey Staff 2006) (Fig. 4.13 and Table 4.4).
- **Ferralitic and pseudo-ferralitic soils**: These are deep colorful soils, especially the B horizon, developed from much altered rocks that have generated 1:1 clay minerals (kaolinite and halloysite), iron oxides (goethite and

Fig. 4.8 Landscapes of the Cordillera of Napo with heavily disturbed vegetation (photo Darwin Sánchez)

Fig. 4.9 Typic Hydrudands soil profile characteristic of the landscapes of the Napo Cordillera (picture Darwin Sánchez)

hematite), and in many cases aluminum oxides (gibbsite). These soils have a base saturation <50% (desaturated) due to the intensity of the rainfall (>2500 mm) which promotes the presence of Udic and Perudic moisture regimes (Zebrowski and Sourdat 1997).

The horizon sequence of the soil profile is A/B/C, with distinct horizons more or less, developed, modified, or rejuvenated. Under the original forest cover, the first 20 cm of the profile are rich in organic matter (>4%). Horizon texture depends on the original material and they might have quartziferous residues (up to 90%). Surface horizons are often poor in clay, but there is no evidence of clay accumulation in the B horizon and no argillic horizons are formed (Zebrowski and Sourdat 1997).

Taking into account the climatic and topographical conditions of the zone, all these soils should be classified as ferralitic, with a Si:Al ratio <2, however, there is a significant proportion of soils with 2:1 clays with a Si:Al ratio >2. Additionally, in these soils, except in the case of rejuvenated by erosion or colluvial soils, base leaching is high, the amount of exchangeable elements is generally <4 cmol kg^{-1}, base saturation, typically <10%, decreases to <1%. The amount of free Fe is high and the free Fe:total Fe ratio ranges

Table 4.3 Description of characteristics of the soil profile classified as Typic Hydrudands (description of Darwin Sánchez)

Horizonts	Depth (cm)	Description
A	0–10	Very dark brown (10YR 2/2) moist color, clayey loam texture, sub-angular blocky and granular structure, bulk density 0.37 g cm^{-3}, strong reaction to NaF, pH 4.6, CEC 5–10 cmol kg^{-1} and base saturation <30%
Bw	10–30	Dark yellowish brown (10YR 5/4) moist color, silty clay loam texture, sub-angular blocky and angular structure, bulk density 0.45 g cm^{-3}, strong reaction to NaF, pH 5.9, CEC 5–10 cmol kg^{-1}, base saturation <30%
2C1	30–45	Yellowish brown (10YR 5/4) moist color, clayey texture, with no structure (massive)
2C2	45+	Light gray (10YR 8/1) moist color, clayey texture, massive structure

Fig. 4.10 Landscape from the Cutucú mountain range without a volcanic ash cover (photo Darwin Sánchez)

from 0.5 to 0.9, the amount of exchangeable Al is exceptionally high, and the pH is generally <5.5, but it and can drop to values <4 (Zebrowski and Sourdat 1997).

The very acid desaturated soils, richer in exchangeable Al, are those which contain the highest amount of 2:1 minerals. These soils are called pseudo-ferralitic, because it is difficult to identify their characteristics and to define their extension in relation to the true ferralitic soils.

The ferralitic soils are represented by those classified as Typic Hapludox (Soil Survey Staff 2006) (Fig. 4.14 and Table 4.5). These are highly evolved old soils with poor fertility and high in exchangeable Al. The oxic horizon is the most important diagnostic property of these soils (CIAT 1983).

On the other hand, the pseudoferralitic soils are classified as Oxic or Typic Dystrudepts (Soil Survey Staff 2006) (Fig. 4.15 and Table 4.6) and are characterized for having an incipient pedogenic development, which has led to the formation of a cambic (Bw) altered horizons. In general, the horizons of this soils have a CEC <24 cmol kg^{-1} in the case of Oxic Dystrudepts and >24 cmol kg^{-1} in the case of the Typic Dystrudepts. The horizon sequence in the profile is as follows: A/B/C type, the first horizon color is brown to dark yellowish brown; the underlying Bw horizon has yellowish brown to yellowish color in the Typic Dystrudepts, and a reddish brown to red color in the Oxic Dystrudepts.

Fig. 4.11 Landscape from the Cóndor mountain range without volcanic ash cover (Missouri Botanical Garden 2007)

Sub-Andean Foothills

This is a transition zone located between the sub-Andean cordilleras and the Peri-Andean hills (unit 4 in Fig. 4.2), with altitudes close to 1000 m o.s.l. at the north, down to about 500 m o.s.l. to the south. The mean annual temperature varies from 21 to 24 °C, and the mean annual precipitation is close to 3000 mm (Sourdat and Winckell 1997).

The strong tectonism and folding during the Andean uplift promoted the presence of foothills made of irregular and angular strongly dissected reliefs, dissected slopes, chevrons, and irregular concave steep slopes having sandy and reddish clayey sedimentary materials. These materials have emerged by the action of ancient erosion processes and differ from the piedmonts of coarse sediments from the north and tuffaceous sandstone and clayey fans from the south (Fig. 4.16) (Sourdat and Winckell 1997).

The soils of this unit are typical tropical soils from areas of high rainfall, with predominantly red (5YR 5/6 to 2.5YR 4/8) and yellow (2.5Y 5/2 to 5Y 6/6) colors according to Fe oxidation status, usually well drained, and clayey loam texture (Zebrowski and Sourdat 1997). Due to the high precipitation, soil base saturation is low (<35%), and soil acidity limits microbial activity in the organic layer (Mejía 1997). For all these conditions, these soils of low fertility and toxic Al content are classified as Oxic or Typic Dystrudepts (Soil Survey Staff 2006) (Fig. 4.17).

In this area, soils classified as Lithic Udorthents can also be found, characterized for having a thin layer of soil and organic material over the rock (Fig. 4.18). These soils were developed on steep slopes and removal of the vegetation cover favors the loss of the surface horizons (O and A) and the outcrop of parental material. For this reason, it is not advisable to use these soils for agriculture or livestock purposes (Añazco et al. 2010; Cañadas 1983).

4.2.2 Peri-Andean Amazonia Reliefs

The Peri-Andean term refers to all areas related to the Andean and Sub-Andean reliefs, particularly to the volcanic nature of the Piedmont materials and the edaphologic features of its soils (Sourdat and Winckell 1997). The strip of these well-represented piedmonts, particularly in the central and south parts, are located here along with the lower hills and fluvial plains (Zebrowski and Sourdat 1997).

Piedmonts

The piedmonts landscapes are established over spread formations localized bellow the Andean and Sub-Andean Reliefs and are characterized for having two essential features: (a) flat summits that have maintained shape even after profound dissection processes; and (b) nature generally feralitic and yellowish color of their soils developed over sandstone or sand from volcanic origin, with the exception of the piedmonts close to the cordillera covered by beige Andisols developed over volcanic ash material (Sourdat and Winckell 1997).

Piedmonts covered with volcanic ash: This unit corresponds to the highest reliefs of the Peri-Andean piedmont

Fig. 4.12 Landscape of the Santiago mountain range showing a pattern of disturbed vegetation (picture Darwin Sánchez)

and represents a broad complex of dissected glacis (Fig. 4.19) extending over Puyo, Palora, and Macas, at altitudes ranging from 1100 to 900 m o.s.l. (unit 5 of Fig. 4.2). These cones were formed by the alluvial contribution of the Pastaza, Palora, and Upano rivers, which additionally had large depositions of volcanic ash from the Tungurahua and Sangay volcanoes. The prevailing weather conditions, with annual mean temperatures of around 20 °C and a mean annual rainfall approaching 5000 mm, are forming factors that have promoted the development of hydrated Andisols classified as Hydrudands (Soil Survey Staff 2006) characterized by their high moisture retention (>100%). Given the Perudic moisture regime of these soils, alteration of primary materials is almost complete, resulting in a mixture of allophane, organic matter, and sesquioxides (Pacheco 2009; Sourdat and Winckell 1997).

The modal profile of this type of soils presents the following horizons sequence: Ap/AB/Bw (Fig. 4.20 and Table 4.7), where the formation of a cambic horizon (Bw) is evident, as an indication of the little pedogenic development which is characteristic of the Andisols.

These are deep soils with good structure, low bulk density (<0.8 g cm^{-3}), and good drainage conditions. However, moderate drainage conditions are evident towards the areas of flat slopes due to the clay content increment of the deeper horizons, which hinders the water movement through the soil profile.

Chemical analysis demonstrated that these soils are acid (pH 5.9), high organic matter content (3.2%), CEC from 5 to 10 cmol kg^{-1} and low base saturation (<35%), conditions which are not favorable to retain essential cations for plant nutrition (Mejía 1997).

Piedmonts over detrital material with no ash cover: These piedmonts correspond to glacis or very wide scatted cones, generally with soft slopes open from the west to the east, between the great courses of the Napo and Pastaza rivers (unit 6, 7, and 8 in Fig. 4.2). The original parental material comes from the Meza formation, which corresponds

Fig. 4.13 Poorly evolved profiles of soils classified as Typic Udorthents (left) and Lithic Udorthents (right) (IEE 2015)

Table 4.4 Description of the characteristics of the soil profile classified as Typic Udorthents (IEE 2015)

Horizons	Depth (cm)	Description
A	0–10	Very dark brown (10YR 2/2) moist color, clayey loam texture, sub-angular blocky structure, pH 5.5, organic matter 1.2%, CEC 18 cmol kg^{-1}, and base saturation 82.2%
C	10–75	Yellowish brown (10YR 5/4) moist color, clay texture, massive structure
R	75+	Rock

to a series of Plio-Pleistocene dissected terraces made of large and medium size clastic deposits derived from highland erosion which, additionally, are partially covered by Quaternary deposits (Sourdat and Winckell 1997). These geological materials developed soils classified as Inceptisols with incipient pedogenic development, which, on spite of the fact that they were formed from yellow, plastic clays rich in gravel quartz, do not show evidence of illuviation processes as a sign of weathering.

Altitudes of these zones, ranging from 900 to 300 m o.s.l., are associated with mean annual temperatures >22 °C, and mean annual rainfall close to 4500 mm in normal years (Sourdat and Winckell 1997). These climatic conditions, especially rainfall, have promoted the development of dystrophic soils (base saturation <50%) corresponding to the great group of Dystrudepts (Soil Survey Staff 2006).

The predominant plant cover of these landscapes corresponds to large areas of grasses for extensive livestock grazing (Fig. 4.21). There are also large areas dedicated to oil palm, palmito, and cocoa cultivation taking advantage of the favorable topographical conditions (ANCUPA 2014; INEC 2013).

Most soil profiles described in these areas present the following horizon sequence: Ap/Bw1/Bw2 (Fig. 4.22 and Table 4.8). These are moderately drained deep soils, meaning that water flows with certain difficulty, due to high clay

Fig. 4.14 Profile of a soil classified as Typic Hapludox (left) located at a very high hilly relief (right) (IEE 2015)

Table 4.5 Description of the characteristics of the profile of a soil classified as Typic Hapludox (IEE 2015)

Horizons	Depth (cm)	Description
A	0–30	Dark yellowish brown (10YR 4/4) moist color, sandy loam texture, sub-angular blocky and granular structure, pH 4.6, organic matter 4.4%, CEC 20 cmol kg^{-1}, base saturation 19.8%
Bo	30–60	Strong brown (7.5YR 5/6) moist color, sandy clay loam texture, sub-angular blocky structure, pH 4.8, organic matter 1.5%, CEC 14 cmol kg^{-1}, base saturation 16.7%
Co	60–100	Olive (5Y 5/6) moist color, silty clay texture, massive structure

content throughout the profile, causing temporary flooding in flat areas.

These are also soils with low base saturation (<35%) and low CEC (5–10 cmol kg^{-1}), medium to low of organic matter content, and a pH ranging from mild to severe acidity, which promotes the presence of Al toxicity problems. These unfertile soils were developed on very humid zones, condition which promotes high leaching of Si and other cations such as Na^+, K^+, Ca^{2+}, and Mg^{2+} (Mejía 1997).

Peri-Andean Hills over Sand Stones, Conglomerates, and Clays

These are the landscapes of the great Amazonia basin composed mainly by hilly massifs located between the break of slope of the Sub-Andean front and the border with Peru, at altitudes ranging from 600 to 250 m o.s.l. (units 9 and 10 of Fig. 4.2). The Piedmont characteristics, strongly marked by granitic, metamorphic, or volcanic origin of the surface materials, are gradually less noticeable, but never absent, in these landscapes justifying the name Peri-Andean given to these units. All the Peri-Andean sedimentary formations are located over the Guyana Shield; they are almost horizontal and have almost no uplifting, which explains the abrupt structural discontinuity and the rupture of slope. The hills have been mapped as a Myo-Pliocene outcroppings when, in fact, the Myo-Pliocene substrate is always present as dissected hills or eroded platforms (Sourdat and Winckell 1997).

Peri-Andean mean annual temperature is >23 °C, and mean annual precipitation range from 2500 up to 4500 mm at the break with the Sub-Andean front. Since there is variation in altitude and rainfall parameters, soil differences are distinguished by the quality of drainage (Sourdat and Winckell 1997).

Fig. 4.15 Profiles of the soils classified as Oxic Dystrudepts (left) (IEE 2015) and Typic Dystrudepts (right, photo Darwin Sánchez)

Table 4.6 Description of the characteristics of the profile of a soil classified as Oxic Dystrudepts (IEE 2015)

Horizons	Depth (cm)	Description
A	0–20	Reddish brown (5YR 4/4) moist color, loamy texture, sub-angular blocky and granular structure, pH 4.5, organic matter 3.8%, CEC 16 cmol kg^{-1}, base saturation 15.4%
Bw	20–50	Reddish brown (5YR 5/4) moist color, clayey loam texture, sub-angular blocky structure, pH 5.0, organic matter 2.6%, CEC 19 cmol kg^{-1}, base saturation 16.4%
C1	50–60	Reddish brown (5YR 5/4) moist color, clayey texture, massive structure
C2	60–110	Yellowish red (5YR 5/6) moist color, clayey texture, massive structure

Eastern low hills: These are the characteristic landforms of the Ecuadorian Amazonia basin, a set of small reliefs with short gradients (10–20 m), convex, with slopes <40%, equal and leveled with each other, for this reason they are known as half oranges (Fig. 4.23). Higher hills are also present, but they do not exceed 50 m of relative gradient, with steeper flanks associated with alignments or massifs where the leveling is not rigorous. A third landscape is a set of hills with rounded uneven tops, generally of tabular shape limited by a small sharp boundary. This latter group is called "sea of surrounding hills" and is located in the middle of water bodies located in the northeastern part of the Ecuadorian Amazonia (unit 9 Fig. 4.2). Cuts of the hills show layers of conglomerates with pebbles and sand on the north side, while in the south side is more frequent the presence clay, sand or pelleted strata (Sourdat and Winckell 1997).

Western Hills: These are mostly unexplored landscapes due to the difficult access; however, photointerpretation

Fig. 4.16 Overview of dissected reliefs covered by natural vegetation (photo Renato Haro)

Fig. 4.17 Profiles of Soil classified as Oxic Dystrudepts (left) and Typic Dystrudepts (right) (Haro and Montúfar 2014)

Fig. 4.18 Vegetation growing in a thin soil layer over rock (Photo by Renato Haro)

shows zones of hills, generally higher than the Eastern Low Hills. The hills are angular, medium to low size, located exclusively in the western part of the Ecuadorian Amazonia, limiting with the Sub-Andean reliefs which have steeper slopes (>40%) (unit 10 Fig. 4.2). This zone also groups the lower flanks of the plateaus and the peripheral reliefs derived from tabular structures, with gentle to rough slopes (Sourdat and Winckell 1997).

Soils of these Peri-Andean hills are the most widespread in the Ecuadorian Amazonia, and are associated with various types of highly acidic red soils (pH < 4.5). They have the following horizon sequence: A/B/C/R, characterized by the dominance of horizons that have passed through intense base and Si leaching, which generated the accumulation of Fe as ferric oxide. The A horizons are similar, but the B horizons are very different from one profile to another because the hilly dissection unexpectedly cuts the layer of original material. These soils are also characterized by the predominance of alteritic horizons with a development that exceeds the depth of the current cuts and never reveal the altered rock. Two general types of soils are presented in this extensive edaphic cover (Custode and Sourdat 1986):

- Lighter, compact red clayey soils, with poorly differentiated shallow horizons on top plinthite containing montmorillonite, illite, vermiculite, and kaolinite, associated with quartz and with <15% oxides. These are desaturated soils(1–20% base saturation) with toxic Al content (Fig. 4.24 and Table 4.9).
- Dark red deeper soils, better organized and drained, with higher clay content consisting of kaolinite, some gibbsite, quartz, and oxides. These are also desaturated soils with a high toxic Al content.

These soils, developed in an Isohyperthermic temperature regime and an Udic to Perudic moisture regime, are classified as Typic or Oxic Dystrudepts or Hapludoxs (Soil Survey Staff 2006). In both cases, the clay content of A horizon is poor, but there is no illuviation evidence. Soil morphology and chemical parameters as base saturation and

Fig. 4.19 Overview of the great piedmont dissected glacis with volcanic ash cover (photo by Jose Merlo)

exchangeable Al indicate that these are low fertility soils. This, along with the relief limitations, reduces drastically the agricultural potential of this zone (Custode and Sourdat 1986).

Alluvial Environment

The unit corresponds to the alluvial valleys (unit 11 Fig. 4.2) with slightly higher relative slope in relation to the rivers, for this reason, the water table usually appears a few centimeters below the surface. Additionally, the predominant flat slopes of these fluvial landscapes make them highly susceptible to flooding. In the lower parts of the Peri-Andean depressions, valleys widen considerably and the entire fluvial environment becomes extensive (Fig. 4.25). The bedrock, mainly composed of boulders transported by the rivers, can only be seen in very deep cuts. The main materials over which the soils have developed are red to gray sandy clay alterites with interbedded grauwaca banks (Sourdat and Winckell 1997). Within the fluvial areas there are swamps and flooded depressions that are distributed along the courses of the rivers.

The soils of the alluvial environment have hydromorphism features, the water table is evident in many of the

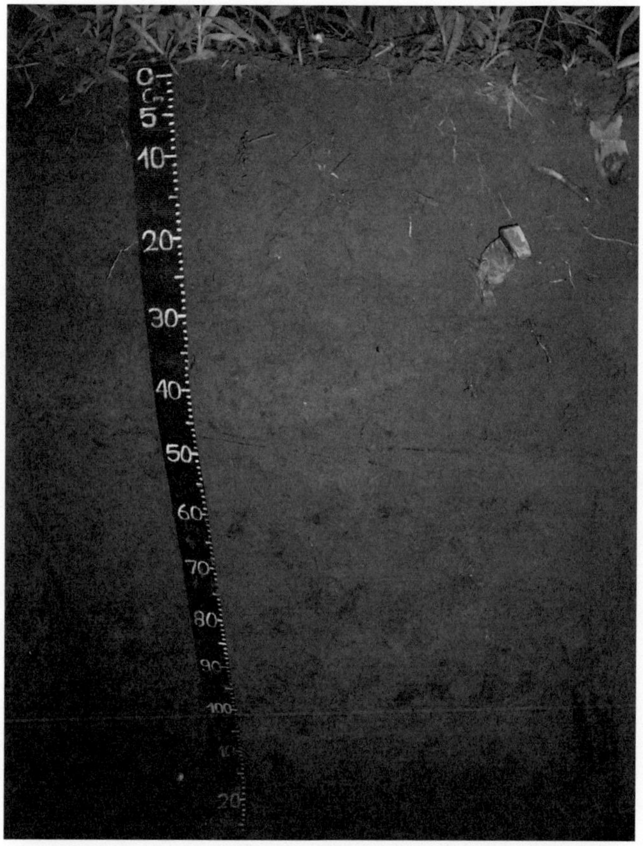

Fig. 4.20 View of a profile of a soil classified as Typic Hydrudands characterized for having high moisture retention (>100%) (Photo by Jose Merlo)

Table 4.7 Description of the profile of a soil classified as Typic Hydrudands (description of Jose Merlo)

Horizons	Depth (cm)	Description
Ap	0–30	Very dark brown (10YR 2/2) moist color, silty loam texture, sub-angular blocky structure, very strong reaction to NaF
AB	30–50	Very dark grayish brown (10YR 3/2) moist color, silty loam texture, sub-angular blocky structure, moderate reaction to NaF
Bw	50–120	Brown (10YR 4/3) moist color, silty clay loam texture, sub-angular blocky structure, slight reaction to NaF

profiles, whereas in those profiles where it does not appear it is evident the presence of redoximorphic features as speckling and gley color. These soils are classified within the Great group of Endoaquepts (Soil Survey Staff 2006), due to the conditions of water soil saturation most of the year.

The representative profile of these soils (Fig. 4.26 and Table 4.10) shows the following horizon sequence: Ap/Bg/Cg. The water table appears at a few centimeters below the surface. In some profiles the water table can be placed below the Cg horizon. Soil pH is moderately

Fig. 4.21 Landscape of the slightly hilly great scattered glacis occupied with grasslands (photo of Jose Merlo)

Fig. 4.22 Profile of a soil classified as Typic Dystrudepts with a clayey loam, yellowish brown color surface horizon and greater clay content deeper in the profile (photo of Jose Merlo)

Table 4.8 Profile description of the soil classified as Typic Dystrudepts (description of Jose Merlo)

Horizons	Depth (cm)	Description
Ap	0–45	Very dark grayish brown (10YR 3/2) moist color, clayey loam texture, sub-angular blocky structure
Bw1	45–85	Dark yellowish brown (10YR 4/4) moist color, clayey loam texture, sub-angular blocky structure
Bw2	85–150	Yellowish brown (10YR 5/6) moist color, silty clay loam to silty clay texture, sub-angular blocky structure

Fig. 4.23 Characteristic landscape of the eastern low hills (photo by Darwin Sánchez)

Fig. 4.24 Typical soil profile of the Peri-Andean landscape (photo by Darwin Sánchez)

Table 4.9 Description of the characteristics of a soil profile classified as Oxic Dystrudepts (description by Darwin Sánchez)

Horizons	Depth (cm)	Description
A	0–10	Dark gray (7.5YR 4/1) moist color, clayey loam texture, sub-angular blockyand granular structure, pH 5.5
Bw1	10–30	Reddish brown (5YR 5/4) moist color, loamy texture, sub-angular blocky structure, pH 5
Bw2	30–75	Yellowing reddish (5YR 5/8) moist color, clayey texture, sub-angular blocky and granular structure, pH 5.3
Bv	75+	Reddish (10R 4/8) as primary color, and light reddish brown (2.5YR 7/3) as secondary mist color, silty clay texture, angular blocky structure and presence of plinthite

Fig. 4.25 Panoramicview of a characteristic flat slope fluvial valley at the Ecuadorian Amazonia alluvial environment (photo of Jose Merlo)

acid (5.5), CEC is <10 cmol kg^{-1} and base saturation is <35%.

Due to the poor drainage conditions, these soils require large investments in civil work to remove the water excess, and need high amounts of organic and mineral amendments to be used for agricultural purposes. These practices would not be profitable and would also have a high environmental impact (Mejía 1997). These landscapes should be preserved without human intervention to maintain the stability of one of the most biodiverse environments of the world (Brondízio et al. 2009).

Fig. 4.26 Soil profile saturated with water typic of the Ecuadorian Amazonia floodplain (photo by Jose Merlo)

Table 4.10 Description of the profile of a soil classified as Typic Endoaquepts (description by Jose Merlo)

Horizons	Depth (cm)	Description
Ap	0–15	Dark brown (10YR 3/3) moist color, silty clay loam texture, and sub-angular blocky structure
Bg	15–33	Olive brown (2.5Y 4/4) moist color, strong brown mottling (7.5YR 5/8), clayey loam texture, sub-angular blocky structure
Cg	33	Olive yellow (2.5Y 6/6) moist color, gray mottling (5Y 6/1), silty clay texture, massive structure

References

Añazco M, Morales M, Palacios W, Vega E, Cuesta A (2010) Sector forestal ecuatoriano: propuestas para una gestión forestal sostenible. Programa regional ECOBONA

ANCUPA (2014) Estadísticas de Palma Aceitera en Ecuador. https://www.google.com/search?q=ancupa&ie=utf-8&oe=utf-8. Accessed 10 diciembre 2014

Bristow C, Hoffstetter R (1977) Léxico Estratigráfico del Ecuador Centre National de la Recherche Scientifique. París, Francia

Brondízio ES, Cak A, Caldas MM, Mena C, Bilsborrow R, Futemma CT, Ludewigs T, Moran EF, Batistella M (2009) Small farmers and deforestation in Amazonia. Amazonia and global change 117–143

Cañadas L (1983) El mapa bioclimático y ecológico del Ecuador. Editores Asociados, Quito, Ecuador

CIAT (1983) Oxisoles y Ultisoles en América Tropical: Distribución, importancia y propiedades físicas. CIAT, Cali, Colombia

Custode E, Sourdat M (1986) Paisajes y suelos de la Amazonía ecuatoriana: entre la conservación y la explotación. Cultura-Revista del Banco Central del Ecuador 8(24):325–337

Espinoza Villar JC, Ronchail J, Guyot JL, Cochonneau G, Naziano F, Lavado W, De Oliveira E, Pombosa R, Vauchel P (2009) Spatio-temporal rainfall variability in the Amazon basin countries (Brazil, Peru, Bolivia, Colombia, and Ecuador). Int J Climatol 29 (11):1574–1594

Haro R, Montúfar C (2014) Reporte del Proyecto Hidroeléctrico Río Santiago. Morona Santiago, MIDENA, Quito, Ecuador

IEE (2015) Perfiles de suelos de Ecuador. Geoportal Instituto Espacial Ecuatoriano, IEE, Quito, Ecuador

INEC (2013) Encuesta de superficie y producción agropecuaria continua Ecuador 2013. Dirección de Estadísticas Agropecuarias y Ambientales, INEC, Quito, Ecuador

IRD, PETROPRODUCCIÓN, IGM (2001) Mapa geológico de la cuenca Oriente. provincias de Morona Santiago, Napo, Pastaza y Sucumbios. Escala 1: 750 000. IRD-Petroproducción, Quito, Ecuador

Letamendi JE, Pérez RM, Huertas AD, Bastardas AP (2007) The Impact of carbonate cements on the reservoir quality in the Napo Fm sandstones (Cretaceous Oriente Basin, Ecuador). Geologica Acta 5(1):89–107

Mejía L (1997) Suelos del Ecuador: Reconocimento general en base a su capacidad-fertilidad y mapa general de clasificación por

Capacidad-Fertilidad de los suelos del Ecuador: Una interpretación básica sobre las características de los suelos del Ecuador Fundación Forestal Juan Manuel Durini; Instituto de la Potasa y el Fósforo (INPOFOS) - Instituto Geográfico Militar (IGM) - Instituto panamericano de Geografía e Historia (IPGH) - Sociedad Ecuatoriana de la Ciencia del Suelo (SECS). Quito, Ecuador

Mena CF, Walsh SJ, Frizzelle BG, Xiaozheng Y, Malanson GP (2011) Land use change on household farms in the Ecuadorian Amazon: design and implementation of an agent-based model. Appl Geogr 31 (1):210–222

MIDENA, IEE, SENPLADES, MAGAP (2013a) Memoria técnica de geopedología del cantón Sucúa del proyecto: Generación de geoinformación para la gestión del territorio a nivel nacional escala 1: 25 000 IEE, Quito, Ecuador

Missouri Botanical Garden (2007) Botanical Exploration of the Cordillera del Cóndor. MOBOT. http://www.mobot.org/MOBOT/research/ecuador/cordillera/images/condor_mirador2.jpg. Accessed 9 de febrero 2015

Pacheco R (2009) El Ecuador: Recursos naturales agrícolas y del medio ambiente. Imprenta Colón, Quito

PRONAREG, ORSTOM (1983a) Mapa morfo-edafológico de la provincia de Morona Santiago (zona sur) y Zamora Chinchipe, escala 1: 500 000. PRONAREG-ORSTOM, Quito, Ecuador

PRONAREG, ORSTOM (1983b) Mapa morfo-edafológico de la provincia del Napo, escala 1: 500 000. PRONAREG-ORSTOM, Quito, Ecuador

Soil Survey Staff (2006) Keys to soil taxonomy, 10th edn. USDA - NRCS, Washington

Sourdat M, Winckell A (1997) Los paisajes de la Amazonía ecuatoriana. In: Winckell A, Zebrowski C, Sourdat M (eds) Los paisajes naturales del Ecuador. Centro Ecuatoriano de Investigación Geográfica (CEDIG) - IPGH (Sección Ecuador) - ORSTOM (Francia) - IGM, Quito, Ecuador

Zebrowski C, Sourdat M (1997) Los factores de la pedogénesis y los suelos en Ecuador. In: Winckell A, Marocco R, Winter T et al (eds) Los paisajes naturales del Ecuador, vol Condiciones generales del medio natural. Centro Ecuatoriano de Investigación Geográfica (CEDIG) - IPGH (Sección Ecuador) - ORSTOM (Francia) - IGM, Quito, Ecuador

5 Soils from the Galapagos Islands

Lorena Lasso and José Espinosa

Abstract

The Galapagos Islands were formed by the rise of magma to the surface trough the faults and fractures of the Nazca Plate, lava flowed freely forming large sheet or coiled blankets, which received pyroclastic contributions from the active volcanos. Even at the moment, the Galapagos archipelago is one of the sites with the greatest volcanic activity in the world. Due to its environmental importance, the Galapagos Islands require special and careful management, for this reason, human activities are allowed only in four islands and in areas that are not included within the National Park. Therefore, soil studies were conducted only in areas with human intervention, where surveys are important for soil management and land evaluation. This chapter presents the information for the Santa Cruz and San Cristobal Islands, because these islands have the largest soil area with human intervention, and for being the islands that have the highest population concentration. The main characteristics of the Galapagos soils come from the alteration of relatively recent pyroclastic deposits in different conditions of altitude, climate, and vegetation. Soil orders present in areas under agricultural use in Santa Cruz Island are Alfisols, Andisols, Entisols, Inceptisols, and Mollisols, while soil orders present in areas under agricultural use in San Cristobal Island are Alfisols, Entisols and Inceptisols.

Keywords

Galapagos islands • Pahoehoe lava • Burned lava • Pyroclastic deposits • Volcanic cones Basaltic rock

5.1 Introduction

The Galapagos Islands are located at 1000 km west of the Ecuadorian coast line and are made up of seven main islands: Isabela, Santa Cruz, Fernandina, Santiago, San Cristobal, Floreana, and Marchena; 14 smaller islands: Spanish, Pinta, Baltra, Santa Fe, Pinzón, Genovesa, Rábida, Seymour Norte, Wolf, Turtle, Bartholomew, Darwin, Daphne Major and South Plaza; 12 additional islands, 64 small islands and 136 rocks, accumulating an area of approximately 8000 km^2 (Ministerio del Ambiente 2014). Only five of these islands are inhabited, while San Cristobal, Santa Cruz, and Isabela are urban parishes, and Floreana and Baltra are rural parishes (Granda and Choez 2013).

The particular climate conditions induced by its location at the equator, and the effect of marine currents of Humboldt and El Niño, have made from the Galapagos Islands a single unique place in the world where singular wildlife that attract the scientists visit and tourists from around the world have developed.

The Galapagos Islands were discovered accidentally by Fray Tomas de Berlanga in 1535; however, there are accounts mentioning that the Inca Tupac Yupanqui made expeditions to the island of fire (Isabela Island) and the outside island (San Cristobal) before Spaniards. The discovery of the Galápagos reported to King Charles V of Spain

L. Lasso (✉)
Ministerio de Agricultura y Ganadería, Quito, Ecuador
e-mail: lore_lasso13@hotmail.com

J. Espinosa
Facultad de Ciencias Agrícolas, Universidad Central del Ecuador, Quito, Ecuador
e-mail: aespinosa@uce.edu.ec

J. Espinosa et al. (eds.), *The Soils of Ecuador*, World Soils Book Series,
https://doi.org/10.1007/978-3-319-25319-0_5

promoted the visit of other Spaniard and their location was spread among seamen. Between 1684 and 1816, the archipelago was frequently visited by pirates, who regarded the islands as site to obtain provisions and water, as well as a hiding place after assaulting Spanish galleons laden with gold. In 1793, James Colnet entertained the possibility of converting the islands in an operation center for whalers (Black 1973).

From their discovery until 1831, the Galapagos Islands were not part of any particular nation. In 1832, the President Juan José Flores authorized an expedition with the mission to take possession of the archipelago making it Ecuadorian territory. Some of the relevant scientific and political events in the history of the Galapagos Islands are presented below (Black 1973):

- Charles Darwin visits the islands in 1835 to complement their studies that led to the theory of evolution and publication of his book "The Origin of Species" in 1859.
- Theodor Wolf, scientist commissioned by the government of Ecuador, visits the islands in 1875 to conduct geological, botanical and zoological studies.
- Rollo Beck organized an expedition sponsored by the California Academy of Sciences in 1906. He arrived accompanied by specialists in geology, entomology, ornithology, botany, zoology, and herpetology, who conducted detailed studies in their respective areas.
- Misael Acosta Solís, Ecuadorian geobotanist, visited the islands in 1935 to collect samples of vegetation and published his observations in the annals of the Central University Ecuador.
- During World War II, the US government established, with government authorization, a naval base at the Baltra Island.
- The Galapagos Islands were declared National Park by the Ecuadorian government in 1959.
- In 1979, UNESCO declared the archipelago as a World Natural Heritage and in 1985 as a Biosphere Reserve (Ishwaran 2012).
- In 2001, the Ecuadorian Government takes control of the human settlements to limit population growth and intervention at the Islands (Congreso Nacional 2001).

The history of the Galapagos Islands soil studies begins in 1962 with the International Scientific Expedition organized by the Department of Geology, University of Ghent, sponsored by the Charles Darwin Foundation and in cooperation with the National Foundation for Scientific Research of Belgium. This expedition collected scientific information including data from geological, geomorphological and pedological studies (Laurelle 1966). The agreement between the Overseas Office of Scientific and Technical Research (ORSTOM), the Galapagos National Institute (INGALA) and the National Agricultural Regionalization Program (PRONAREG) allowed the conduction of thematic natural resources studies at the Galapagos Islands in 1987 (PRONAREG 1987). Finally, the National Program for Information and Rural Land Management and Technological Infrastructure (SIGTIERRAS) conducted comprehensive thematic studies for land valuation of the main islands in 2011 (SIGTIERRAS 2011).

Due to its preservation importance, the Galapagos Islands require special and careful management, for this reason, human activities are allowed only in four islands and in areas that are not included within the National Park.

The SIGTIERRAS studies were conducted mainly in areas with human intervention, where surveys are important for soil management and land evaluation. This chapter presents the information for the San Cristobal and Santa Cruz Islands, because these islands have the largest soil area with human intervention (Fig. 5.1), and for being the islands that have the highest population concentration. San Cristobal has 30% and Santa Cruz 61% of the Galapagos population (Granda and Choez 2013). The San Cristobal Island has 567 km^2, and about 10% of this area is suitable for agricultural use, the remainders are populated areas or areas included in the National Park. The Santa Cruz Island has 988 km^2, and about 11% of the surface of the island is suitable for agriculture and the remainders are populated areas or are also within the National Park (SIGTIERRAS 2011).

Even at the moment, the Galapagos archipelago is one of the sites with the greatest volcanic activity in the world. However, the Santa Cruz and San Cristobal islands, being the oldest, show only remnants of volcanoes that no longer have volcanic activity.

5.2 Origin of the Islands

The Galapagos Islands were formed 20 million years ago from the spring of magma, known as the Galapagos hot spot, located in the mantle over the upper crust between the Cocos and Nazca plates. The raising magma came in contact with the earth crust promoting the formation of submerged and surfaced volcanoes, which originated the Carnegie ridge and then the Galapagos island platform (Fig. 5.2). The current geomorphological shape of the islands was formed by the rise of magma to the surface trough the faults and fractures of the Nazca Plate (Sallarés et al. 2009). Even at the moment, the Galapagos archipelago is one of the sites with the greatest volcanic activity in the world. However, the Santa Cruz and San Cristobal islands, being the oldest, show only remnants of volcanoes that no longer have volcanic activity (Acosta Solís 1979).

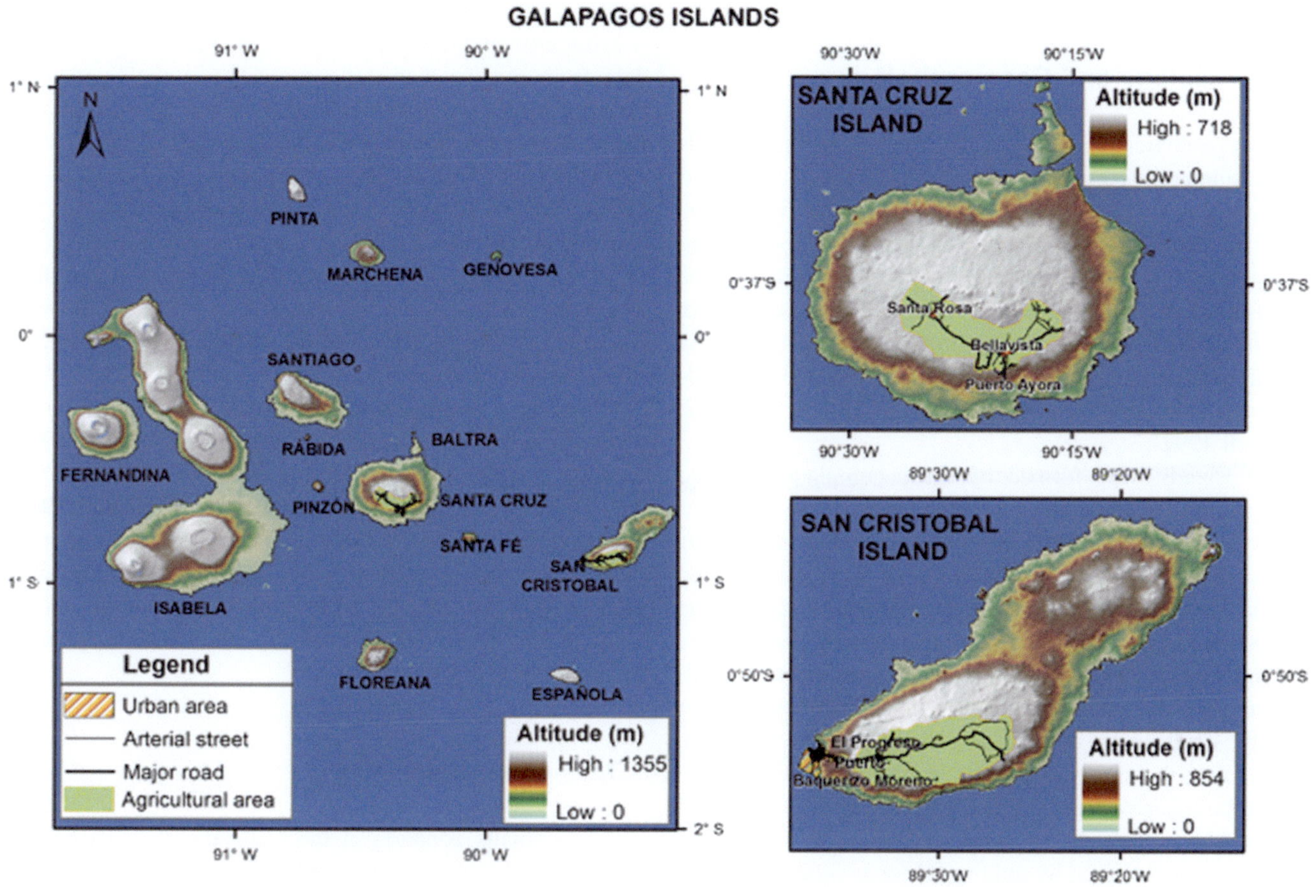

Fig. 5.1 Location and agricultural areas of Santa Cruz and San Cristobal islands (SIGTIERRAS 2011)

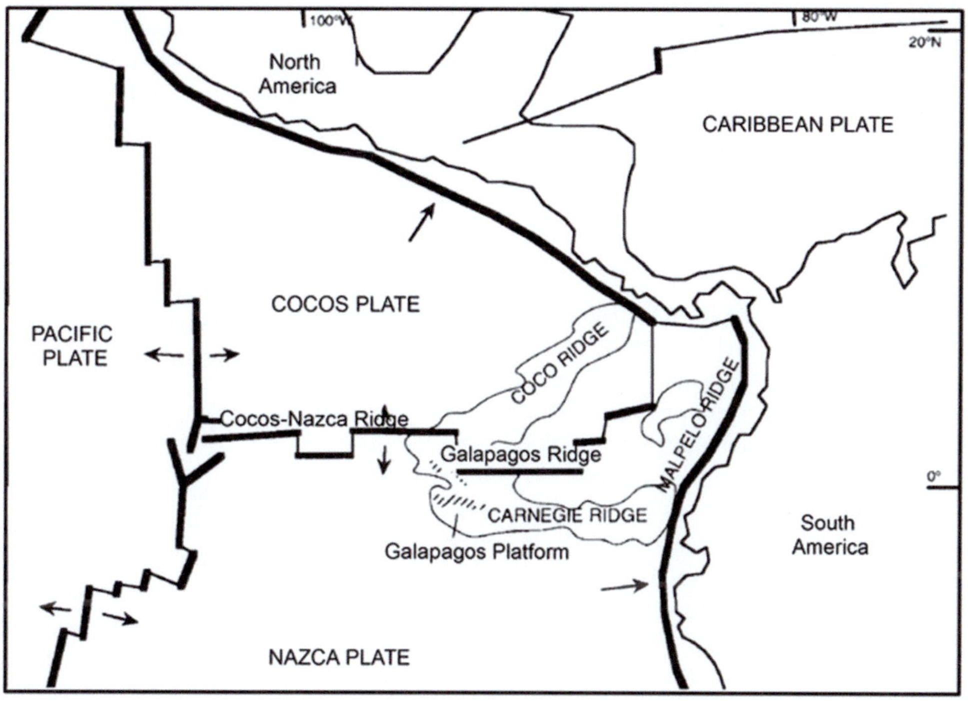

Fig. 5.2 Oceanic plates and the Galápagos Archipelago (Winckell 1997)

5.3 Soil Forming Factors

5.3.1 Geology and Geomorphology

The two geologic units that originated the Galapagos Islands are the submarine basaltic lavas raised from the platform and the young volcanoes built from basaltic and pyroclastic materials (Black 1973). The rocks of the archipelago are almost entirely volcanic and <1% is sedimentary rocks. The islands of Santa Cruz and San Cristobal are made from alkaline basalts rich in magnesium of aphanitic and porphyritic plagioclase nature associated with interbedded sediments of submarine origin (Bristow and Hoffstetter 1977; Winckell 1997).

The islands were formed by the accumulation of two types of lava, one that flowed freely forming large sheet or coiled blankets (pahoehoe lava) and lava that, due to its viscosity, was ejected in blocks which then were stirred forming acute and sharp edges that obstruct the entry into the islands interior (burned or AA lava) (Black 1973). The aphanitic basalt, very coherent and somehow porous, weathers slowly, while the pyroclastic projections weather quickly. The blocks of AA lava have a faster alteration process than the sheets of pahoehoe lava (Winckell 1997).

The landscape of the eastern half of the San Cristobal Island and northeastern and southern parts of the Santa Cruz Island are characterized by mild to moderate reliefs, blankets of basaltic flows of partial submarine origin, and small cones of slag crests that do not exhibit the high volcanic buildings of the other islands. These areas are characterized for presenting shield volcanoes, partially eroded, with calderas difficult to identify. Both islands have an extensive basal part with gentle slopes which increases gradually with the altitude to finally fuse with the flat or gently rolling relieve from the top (Winckell 1997).

5.3.2 Climate

The Galapagos Islands have the same climatic conditions as the continental Coastal Region of Ecuador, however, they have the influence of antagonist marine currents that change the climate: the warm current of El Niño from the north, the cold currents of Humboldt from the South, and Cromwell from the east (Chiriboga et al. 2006; Winckell 1997). For this reason, the islands have two distinct seasons, the warm season from January to May and the cold season from June to December (Chiriboga et al. 2006).

On the other hand, the rainfall pattern of the islands is influenced by the altitude. The upper sections, located over 200 m a.s.l., receive >400 mm annual rainfall, whereas the lowlands, at sea level, a precipitation of 200 mm in the warm season and lower amounts during the cold season. The data from the weather stations located at the Santa Cruz Island show that rainfall is low at sea level, but precipitation increases on the slopes exposed to moisture-laden winds (Table 5.1) (Trueman and d'Ozouville 2010; Winckell 1997). The oscillation of the Humboldt Current has two extremes, known as El Niño y La Niña, which periodically affect the climate of the Islands. El Niño can appear from December on increasing the temperature and rainfall, while La Ñina significantly reduces the same climate parameters (Trueman and d'Ozouville 2010).

There is a temperature gradient at the Islands with a pattern that decreases the temperature in 0.65 °C per 100 m increase in altitude, condition that contrasts with the 0.3 °C per 100 m prevalent in continental Ecuador (Huttel 1986). However, this data cannot be extrapolated to the higher parts of all islands because there is not enough supporting information. At sites around 1800 m a.s.l. the average annual temperature would be 0 °C if the described temperature gradient holds (Huttel 1997).

5.3.3 Vegetation

The bioclimatic zonation of the Galapagos Islands demonstrates that there is a close relationship between climatic zones and vegetation type. These bioclimatic zones of the Islands, properly described by Black (1973) and Huttel (1986), are presented bellow and in Fig. 5.3.

Coastal Zone (Hidrohalophytic Edge)

Altitude 0–10 m a.s.l., halophytic plants growing exposed to sea water and very saline soils: mangroves (Rhizofora mangle), arrayancillos (Maytenus octogona) manzanillos (Manchineel) and small succulent plants. This zone consists of rocky areas and sandy beaches.

Dry Zone (Beach or Dry Strip)

Altitude 10–100 m a.s.l., xerophyte vegetation adapted to rocky conditions and lack of moisture: pear cactus (Opuntia galapageia), cactus (Jasminocereus thouarsii), hawthorns (Scutia pauciflora), palosanto (Bursera graveolens) and thorny bushes.

Transitional Zone (Zone of "pegapega")

Altitude 100–180 m a.s.l., xerophyte and mesophyte vegetation from arid and humid climates. Associations of carob (Acacia insulae-iacobi), palo santo (Bursera graveolens), cactus (Opuntia galapageia), cat's claw (Zanthoxylum fagara), ferns (Elaphoglossum firmum), and epiphytes.

Table 5.1 Average temperature and precipitation reported by the two weather stations at the Santa Cruz Island (INAMHI 2014)

Season of the year	Puerto Ayora station (at sea level)		Bellavista station (200 m a.s.l.)	
	Temperature (°C)	Precipitation (mm)	Temperature (°C)	Precipitation (mm)
Warm (January to May)	26.2	274.7	24.9	696.3
Cold (June to December)	22.5	79.4	21.2	497.3

Fig. 5.3 Typical vegetation from the Santa Cruz and San Cristobal Islands: **a** Vegetation from the coastal zone; **b** Transition zone vegetation; **c** Scalesia zone; **d** Natural pastures and grasses zone (Photos Carla Lasso)

Scalesia Zone (Humid Region)

Altitude 180–400 m a.s.l., mesophyte vegetation of epiphyte plants, ferns, lianas, but here predominate Scalesias (Scalesia affinis), miconias (Miconia robinsoniana) and guayabillos (Psidium galapageium). Land in this zone is also utilized to grow coffee (Coffea arabica) associated with plantain (Musa paradisiaca) for shade, and natural pastures utilized for cattle.

Superior Floor (Miconias Zone, Ferns)

Altitude >400 m a.s.l. From 400 to 550 m a.s.l. the predominant vegetation is made of cacaotillos (Miconia robinsoniana), tree ferns (Cyathea weatherbyana) and orchids. Above 550 m a.s.l. the vegetation is characterized by the presence of grasses such as Paspalum pinicillatum, Paspalum longepedunculatum, Panicum colonum, Calamagrostis pumila, vegetal material similar to those plants that grow in the Andean paramo.

5.4 Soils from the Santa Cruz Island

Studies conducted by the PRONAREG-ORSTOM-INGALA group identified three different zones at Santa Cruz (PRONAREG 1987): (a) High, very humid and temperate zone, (b) Intermediate wet and warm zone, and (c) Bottom dry, very dry, and warm zone. These areas were defined based on the relationship between altitude, climate and vegetation, which are the determinant factors for soil formation.

The main characteristics of the Galapagos soils come from the alteration of relatively recent pyroclastic deposits in different conditions of altitude, climate, and vegetation. Figure 5.4 shows a photograph of a typical landscape and Fig. 5.5 shows the Santa Cruz Island soil map. A detailed discussion of the soils of each of these zones is as follows.

Fig. 5.4 Santa Cruz Island landscape characterized for the presence of well-preserved volcanic constructions and rolling plains (Photo Gustavo Tapia)

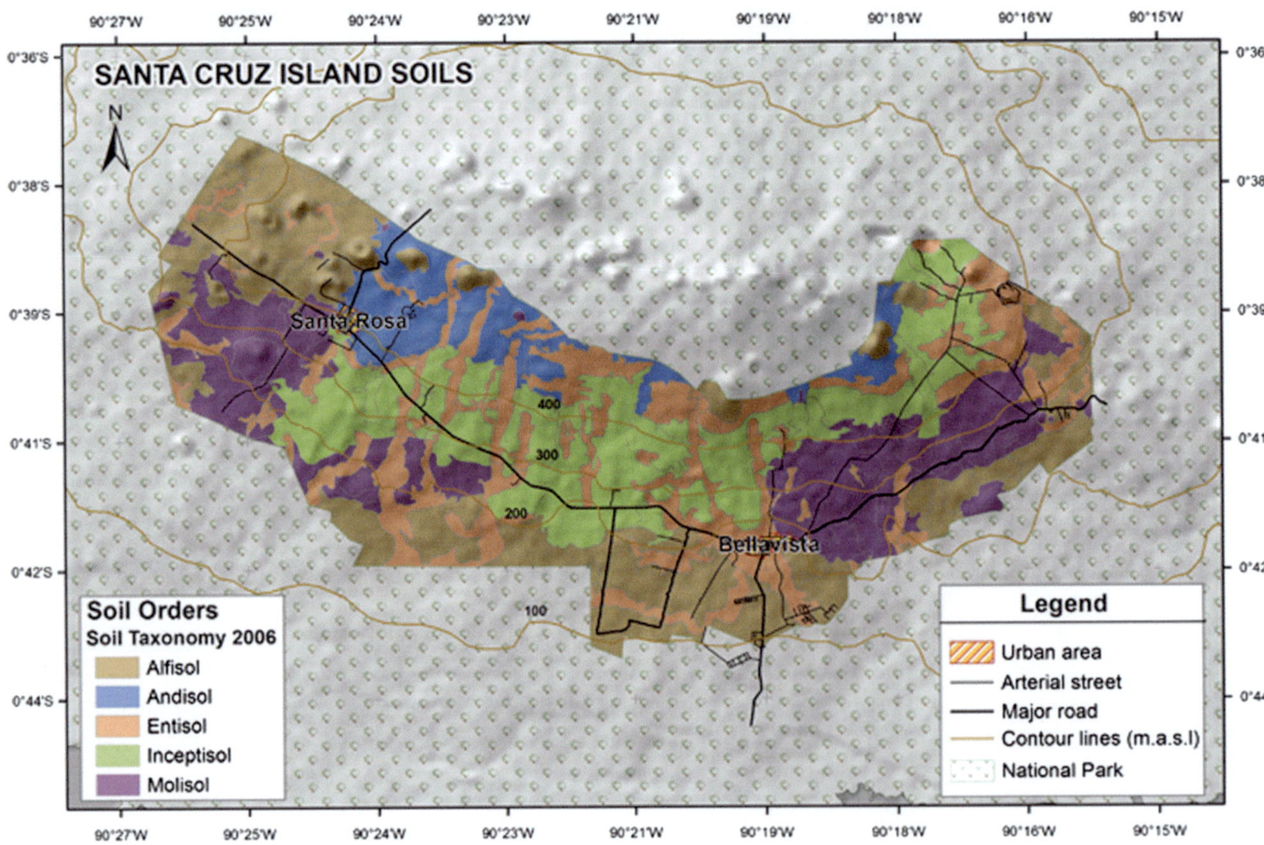

Fig. 5.5 Santa Cruz Island soil map (PRONAREG 1987; SIGTIERRAS 2011)

5.4.1 High, Very Humid and Temperate Zone

This zone of the Island soils is located at the miconias area and ascends towards the grasses area, receives a precipitation ranging from 800 to 1800 mm and reaches temperatures ranging from 21 to 19 °C (SIGTIERRAS 2011). The landscape is characterized by the presence of small well-preserved volcanic cones and cones in a clear destruction process, flat or slightly undulating reliefs with accumulation of ash and pyroclastic material, evidence of runoff channels with rocky outcrops, and colluvial deposits with sandy-silty-gravel materials (Fig. 5.6). These landscapes are in many respects similar to the Andean paramos (Winckell 1997).

In this zone, the influence of basaltic rock disappears and fragments of basaltic rock within the pyroclastic deposits can be observed. This change is evident in the soil characteristics which are slightly acidic to acidic, and have low base saturation, constant organic carbon content through the profile and the presence of amorphous short range clays (Table 5.2) (Laurelle 1966).

PRONAREG (1987) classifies the soils in this area also as Andisols having the following characteristics: yellowish brown color, silty loam texture, moderate depth, moderate reaction to NaF, bulk density <0.9 g cm^{-3}, and slightly acidic pH. Soils on the slopes of the volcanic cones are classified as Typic Hapludands, while soils at the colluvial areas with presence of coarse rock fragments are classified as Lithic Hapludands.

5.4.2 Intermediate Wet and Warm Zone

The Intermediate zone is located from 200 to 400 m a.s.l., with rainfall ranging from 800 to 1000 mm and temperatures from 23 to 21 °C (SIGTIERRAS 2011). The landscape from this zone corresponds to large slopes formed by poorly dissected lava surfaces and runoff channels where colluvio alluvial deposits accumulated (Winckell 1997). Due to its location, the lavas of this area have undergone major alterations, and the rocky outcrops have worn surfaces with partial coating of pyroclastic material (Fig. 5.7) (SIGTIERRAS 2011).

Fig. 5.6 Soils from the high, very humid and temperate zone (Photo Carla Lasso)

Two types of soils are identified in this zone, both characterized for having an A/B/C horizon sequence, but one of them has a developed B horizon (Bw) characteristic of an Inceptisols, while the other has a B horizon enriched with clay (Bt) characteristic of an Alfisols (Laurelle 1966).

On the other hand, studies carried out by PRONAREG in 1987 noticed that the predominant soils from this zone are Typic Dystrudepts (order Inceptisols until 1987). These are reddish brown soils, with silty clay loam texture, slight reaction to NaF, moderately deep, <50% base saturation, and slightly acid. This type of soil is largely present on the areas with slight slopes, while soils developed over the colluvial deposits and the rocky outcrops from lava fluxes are classified as Lithic Dystrudepts.

The influence of the basaltic rock on edaphic development is evident in soil pH and base saturation, but the presence of pyroclastic material is responsible for the texture (Laurelle 1966). Additionally, the PRONAREG study (1987) also found Alfisols represented by Typic Hapludalfs and Mollisols, mostly Typic Argiudolls, coinciding with the soil characterization made by the Geopedologic Mission of Ghent (Table 5.3).

5.4.3 Bottom Dry and Very Dry Warm Zone

The zone extends from 100 to 200 m a.s.l. has a rainfall pattern varying from 600 to 800 mm, a mean temperature of 24 °C (SIGTIERRAS 2011), undulating surfaces product of poorly weathered Pahoehoe lava flows, and is colonized by a shrubby vegetation (Winckell 1997).

The representative soils of this zone are Alfisols, followed in importance by Mollisols and Entisols. The Alfisols, developed on corrugated lava surfaces, correspond to the Vertic Haplustalfs subgroup, and are characterized by the reddish brown color, high clay content, base saturation >50%, CEC >30 cmolc kg^{-1}, and pH 7.2 (PRONAREG 1987). Among the Mollisols, the predominant subgroup is Lithic Argiustolls that appear on the remaining old cones present on the undulating surfaces (Fig. 5.8 and Table 5.4).

Among the Entisols, the most prominent are those of the Typic Ustorthents subgroup, located on the moderate slope flanks with presence of abundant coarse fragments and rock outcrops. These soils are characterized by an A/C horizon

Table 5.2 Soil profile characteristics of an Andisol characterized by the Geopedologic Mission of the University of Ghent (Laurelle 1966)

Horizons	Depth (cm)	Description
A11	0–6	Very dark brown (10YR 2/2) color, loamy texture, very friable to friable weak granular structure, pH 6.0, CEC 66.6 cmolc kg^{-1}, base saturation 4.3%
A12	6–20	Dark brown (10YR 3/3) color, clayey texture, very friable massive structure, pH 6.2, CEC 68.6 cmolc kg^{-1}, base saturation 5.3%
C	20–90	Yellowish brown (10YR 3/4) color, very friable to friable massive structure, presence of abundant tuffs, pH 6.0, CEC 47.5 cmolc kg^{-1}, very low base saturation

Fig. 5.7 Intermediate wet and warm zone showing the scalesias growing over rock (Photo Carla Lasso)

Table 5.3 Soil profile characteristics of an Argiudolls from the Intermediate zone characterized by the Geopedologic Mission of the University of Ghent (Laurelle 1966)

Horizons	Depth (cm)	Description
A11	0–5	Brownish black (7.5YR 3/2) color, silty loam texture, very friable to friable strong granular structure, pH 7.4, CEC 50 cmolc kg^{-1}, base saturation 72.2%
A12	5–13	Brownish black (7.5YR 3/2) color, silty loam texture, moderately firm granular structure with few rock fragments, pH 7.0, CEC 50.6 cmolc kg^{-1}, base saturation 57.5%
Bt	13–40	Yellowish brown (10YR 4/3) color, silty loam texture, blocky to subangular structure with many small and medium sized rock fragments, clay coatings in pores, pH 6.8, CEC 35.9 cmolc kg^{-1}, base saturation 74.4%
C	40–75	Reddish brown (5YR 4/4) color, tuffaceous material mixed with ashes, unstructured, pH 6.5, CEC 36.8 cmolc kg^{-1}, base saturation 68.2%

sequence mainly developed on altered of basaltic rocks (PRONAREG 1987). The pH and base saturation range from 6.5 to 7.5 and 70 to 85%, respectively, and there is evidence of clay liners, indicating the presence of illuviation processes. The clay fraction contains halloysite and smectite, with lower amounts of feldspars (Stoops 2014).

Fig. 5.8 Soil profile with presence of abundant rock fragments classified as Lithic Argiustolls (Photo Gustavo Tapia)

Table 5.4 Soil profile characteristics of an Argiudolls from the Bottom zone described by the Geopedologic Mission of the University of Ghent (Laurelle 1966)

Horizons	Depth (cm)	Description
A11	0–20	Brownish black (5YR 2/2) color, clayey texture, strong granular structure with small pore coatings, pH 7.1, CEC 59.2 cmolc kg^{-1}, base saturation 78.3%
A12	20–36	Dark reddish brown (5YR 3/2) color, clayey texture, granular structure with many fragments of weathered rock, pH 6.8, CEC 31.1 cmolc kg^{-1}, base saturation 64%
Bt	36–70	Brownish black (10YR 2/2) color, clayey texture, weak blocky structure, clay coatings through and within of the peds, difficult root penetration, pH 6.7, CEC 50.9 cmolc kg^{-1}, base saturation 89%
R	70+	Unchanged parent material

Table 5.5 Soil orders present in areas under agricultural use in Santa Cruz Island (SIGTIERRAS 2011; Soil Survey Staff 2006)

Orders	%
Alfisols	26.0
Andisols	8.1
Entisols	23.1
Inceptisols	22.9
Mollisols	19.9

As noted above, the areas where soil surveys were performed are those where agricultural activities are conducted, which is only a small portion of the total area of the Island. Conservation regulations do not allow agricultural expansion in the Galapagos Islands. Table 5.5 shows the percentage of each soil order present on the land occupied by agricultural activity at the Santa Cruz Island.

5.5 Soils from the San Cristobal Island

Agricultural activities occupy 17.7% of the total area of the San Cristobal Island, and inside of this agricultural area four zones can be identified: (a) High wet and very wet temperate zone, (b) Intermediate wet and warm zone, (c) Low dry and warm zone, and (d) Low very dry to arid warm zone, according to the relationship among soil forming factors like altitude, climate, and vegetation.

The landscape on top of the island consists of rolling plains with the presence of well-preserved slag constructions of strong and moderate reliefs. The lower part has gentle interior flanks and colluvio alluvial valleys. The western part of the island consists of an ancient shield volcano with a disappearing caldera without recent activity. The soils from this area present a more advanced state of morphological evolution (Winckell 1997).

The San Cristobal Island is older than the Santa Cruz Island; therefore, the pedogenic processes are much more advanced (Stoops 2014). The soils are highly cohesive red soils that are found from the very humid highlands to the dry lowlands (Franz 1980). Alfisols are the predominant soils in most landscapes, but there are also small areas with Entisols, mainly in the colluvial alluvial valleys over sandy-silty-gravel deposits (Winckell 1997).

5.5.1 High Wet and Very Wet Temperate Zone

This zone is located over the 410 m a.s.l., receives >1200 mm rainfall and the mean temperature varies from 21 to 18 °C (SIGTIERRAS 2011). This zone is characterized for having undulating surfaces containing well-preserved small conical constructions formed by abundant projections of slag and lapilli that cover the surfaces (Winckell 1997). Mineralogical studies from this zone came upon high contents of gibbsite, in places also hematite, associated with kaolinite (Adelinet et al. 2008). The upper part of the San Cristobal Island, unlike other islands, has high atmospheric humidity and high rainfall, climatic conditions that are evident in the cover vegetation mainly made of Miconias (Miconia robinsoniana) (Winckell 1997).

According to the studies conducted by PRONAREG (1987), the soils in this zone, developed over basaltic rock, have a reddish color, clayey texture, pH 5.3, CEC <24 cmolc kg^{-1}, and base saturation <50%. According to these features, the soils are Alfisols, particularly Typic Rhodudalfs. These soils are present in the scoriaceous cones and in surfaces with variable dissection where there is well weathered pyroclastic material. The study also identified, in a lesser extent, Oxic Dystrudepts (Inceptisols) , characterized for having silty clay loam texture, presence of gravel, and the same chemical characteristics of soils present on the rolling surfaces of this zone.

In the soil profile description shown in Table 5.6 it is evident the presence of umbric and argillic horizons as well as andic characteristics and a bulk density >1 g cm^{-3} (ISRIC 1993).

5.5.2 Intermediate Wet and Warm Zone

The Intermediate zone spreads from 300 to 410 m a.s.l., gets 800–1200 mm average annual rainfall and has temperatures varying from 21 to 22 °C (SIGTIERRAS 2011). This zone is characterized for having flat to slightly wavy surfaces with the presence of watersheds that constitute the drainage network of the water excess during the rainy season. These drainage outlets are characterized by strong notches inherited from a wetter period when significant water movement produced excavations of magnitude (Winckell 1997). The soils in this zone are classified as Alfisols with vertic properties (Vertic Hapludalfs) developed on flat and slightly undulating surfaces, characterized by having reddish brown color, clayey texture, presence of cracks, base saturation 50%, CEC <24 cmolc kg^{-1}, and pH 6.2 (PRONAREG 1987).

5.5.3 Low Dry and Warm Zone

This zone is expands from 200 to 300 m a.s.l., has an annual rainfall ranging from 600 to 1000 mm and temperatures from 22 to 23 °C (SIGTIERRAS 2011). The landscape

Table 5.6 Profile characteristics described by Kauffman and del Posso (ISRIC 1993) from a soil classified as Typic Rhodudalfs (Soil Survey Staff 2006)

Horizons	Depth (cm)	Description
Ah	0–12	Dark reddish brown (5YR 3/4) color, clayey texture, fine subangular blocky structure, slightly sticky and slightly plastic, many fine tubular pores, pH 4.4, CEC 30.9 cmolc kg^{-1}, base saturation 5%
Bts1	12–95	Dark reddish brown (2.5YR 3/6), silty clay texture, fine to medium subangular blocky structure, slightly sticky and slightly plastic, presence of clay and sesquioxides cutans, pH 4.8, CEC 20.3 cmolc kg^{-1}, base saturation 1%
Bts2	95–142	Dark reddish brown (2.5YR 3/6) color, silty loam texture, fine subangular to blocky structure, slightly sticky and slightly plastic, very friable, clay and sesquioxides cutans, pH 4.7, CEC 20.1 cmolc kg^{-1}, base saturation 1%
C	142–200	Dull reddish brown (5YR 4/4) color, silty loam texture, very friable massive porous structure, presence of heavily weathered basalt and pumice fragments, pH 4.8

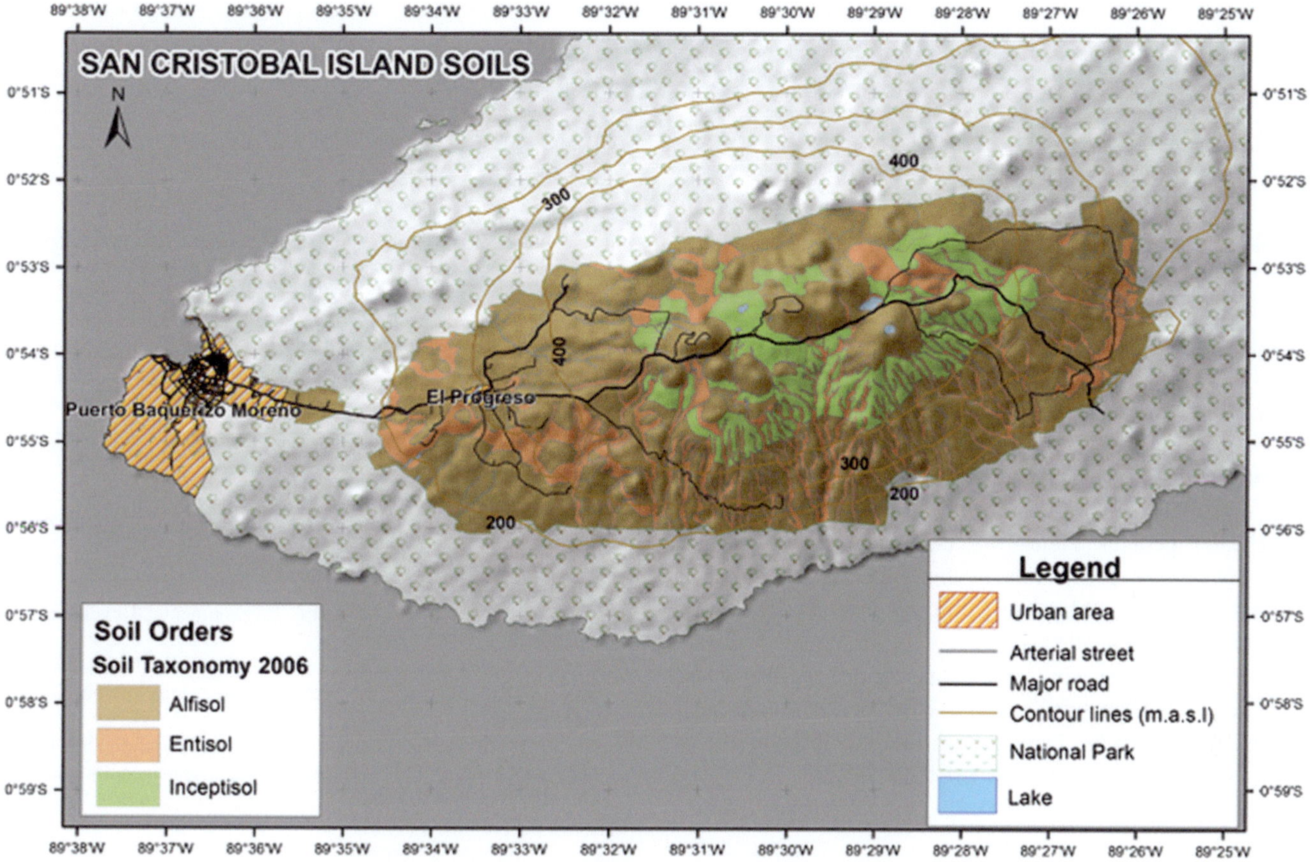

Fig. 5.9 Soil orders present in areas under agricultural use in San Cristobal Island (PRONAREG 1987; SIGTIERRAS 2011)

Table 5.7 Profile characteristics described by Kauffman and del Posso (ISRIC 1993) from a soil classified as Vertic Haplustalfs (Soil Survey Staff 2006)

Horizons	Depth (cm)	Description
Ah	0–10	Dark reddish brown (5YR 3/2) color, clayey texture, blocky to subangular structure with few very weathered basalt fragments, sticky and plastic, pH 6.5, CEC 50.2 cmolc kg^{-1}, base saturation 96%
BA	10–30	Dark reddish brown (5YR 3/4) color, clayey texture, blocky structure with few medium highly weathered basaltic fragment, sticky and plastic, pH 6.3, CEC 31.3 cmolc kg^{-1}, base saturation 87%
Bt	30–67	Reddish brown (5YR 4/6) color, clayey texture, blocky angular structure with fine to medium rock fragments, sticky and plastic, pH 5.4, CEC 37.7 cmolc kg^{-1}, base saturation 78%
C	67–150	Bright brown (7.5YR 5/6) color, sandy clay loam texture, very friable porous massive structure with abundant highly weathered basalt fragments, pH 6.1, and CEC of 44.8 cmolc kg^{-1}, base saturation 90%

consists of gentle slopes, and plains slightly inclined and dissected by colluvial alluvial valleys that come from the upper parts (Winckell 1997). Representative soils from this zone are classified as Vertic Haplustalfs (Alfisols), having a reddish brown color, silty clay loam to clayey texture, slightly acidic pH, base saturation >50%, and CEC 24 cmolc kg^{-1}. These soils are deep, usually cracked due to the presence of smectites (Adelinet et al. 2008). The profile presented in Table 5.7 shows the vertic features and the argillic horizon of these type of soils (ISRIC 1993).

Table 5.8 Soil orders present in areas under agricultural use in San Cristobal Island (SIGTIERRAS 2011; Soil Survey Staff 2006)

Soil orders	%
Alfisols	65.2
Entisols	20.8
Inceptisols	14.0

5.5.4 Low Very Dry to Arid Warm Zone

This zone is located bellow 100 m a.s.l., with rainfall <600 mm and temperatures ranging from 23 to 25 °C (SIGTIERRAS 2011). The landscape is characterized by gentle slightly dissected slopes formed from ancient volcanic emissions which were colonized by shrubs adapted to dry and arid conditions. The weathering was poor due to climate, therefore, rocky outcrops and more clayey, reddish, rich in gravel and pebbles facies are observed on the surface (Winckell 1997). Most soils in this zone are classified as Alfisols (Fig. 5.9), subgroup Lithic Haplustalfs, characterized for having a shallow contact with the parent material, reddish brown color, silty clay loam texture, slightly acidic pH, base saturation >50% and a CEC of 24 cmolc kg^{-1} (PRONAREG 1987). Table 5.8 shows the distribution of agricultural soils from the San Cristobal Island.

References

Acosta Solís M (1979) Galápagos y su naturaleza: Geografía, ecología y conservación. IPGH, Quito

Adelinet M, Fortin J, d'Ozouville N, Violette S (2008) The relationship between hydrodynamics properties and weathering of soils derived from volcanic rocks, Galapagos Islands (Ecuador). Environ Geol 56 (1):45–58

Black J (1973) Galápagos: archipiélago del Ecuador. Fundación Charles Darwin/World Wild Fund, Quito

Bristow C, Hoffstetter R (1977) Léxico estratigráfico del Ecuador. París

Chiriboga R, Fonseca B, Maignan S (2006) Zonificación agroecológica de las zonas agropecuarias en relación con las especies invasoras. In: INGALA (ed) Proyecto ECU/00/G31 Especies invasoras de las Galápagos. MAE, INGALA, UNDP

Congreso Nacional (2001) Ley Especial para la Provincia de Galápagos. Resolución Legislativa No. 22-058, publicada en Registro Oficial 280. Registro Oficial, Quito, Ecuador

Franz H (1980) Old soils and land surfaces on the Galápagos Islands. Geo Journal 4(2):182–184

Granda M, Choez G (2013) Población y migración en Galápagos. Informe Galápagos 2011–2012. DPNG-CGREGE-FCD-GC, Puerto Ayora

Huttel C (1986) Zonificación Bioclimatológica y Formaciones Vegetales en las Islas Galápagos. Revista Cultura, Banco Central del Ecuador, Quito

Huttel C (1997) Las grandes regiones climáticas y sus formaciones vegetales naturales In: Winckell A, Marocco R, Winter T et al (eds) Paisajes naturales del Ecuador: las condiciones generales del medio natural. CEDIG - IPGH - IRD - IGM, Quito

INAMHI (2014) Anuario Meteorológico Estaciones Puerto Ayora y Bellavista Santa Cruz INAMHI. Puerto Ayora, Ecuador

Ishwaran N (2012) Science in intergovernmental environmental relations: 40 years of UNESCO's Man and the Biosphere (MAB) Programme and its future. Environmental Development 1 (1):91–101

ISRIC (1993) Soil Reference Profiles of Ecuador: Field and Analytical Data. ISRIC (International Soil Reference Information Centre), Wageningen

Laurelle J (1966) Study of a soil sequence on Indefatigable Island. Paper presented at the Symposia of the Galápagos International Scientific Project, California

Ministerio del Ambiente (2014) Parque Nacional Galápagos. Ministerio del Ambiente. http://www.ambiente.gob.ec/parque-nacional-galapagos/. Accessed 15 de Diciembre 2014

PRONAREG (1987) Inventario cartográfico de los recursos naturales, geomorfología, vegetación y biofísicos de las Islas San Cristóbal y Santa Cruz. PRONAREG - INGALA, Quito

Sallarés V, Charvis P, Calahorrano A (2009) Naturaleza y formación de la Provincia volcánica de Galápagos. Geología y geofísica marina y terrestre del Ecuador desde la Costa Continental hasta las Islas Galápagos. Guayaquil

SIGTIERRAS (2011) Cartografía temática de los cantones: San Cristobal y Santa Cruz. MAGAP, Quito

Soil Survey Staff (2006) Keys to soil taxonomy, 10th edn. USDA - NRCS, Washington

Stoops G (2014) Soils and paleosoils of the Galápagos Islands: What we know and what we don't know. A meta-analysis University of Hawai Press, Hawai

Trueman M, d'Ozouville N (2010) Characterizing the Galápagos terrestrial climate in the face of global climate change. Puerto Ayora, Isla Santa Cruz. Fundación Charles Darwin, Quito

Winckell A (1997) Los paisajes de las Islas Galápagos. In: Winckell A, Zebrowski C, Sourdat M (eds) Los paisajes naturales del Ecuador: Las regiones y paisajes del Ecuador. CEDIG - IPGH - IRD, Quito

6 Agricultural Land Use

José Espinosa and Julio Moreno

Abstract

Population growth, land pressure, and the demand for food and exportable products are factors that promoted the expansion of the country agricultural frontier. The area under cultivation changed from 1´766,000 ha in 1954 to 2´470,000 ha in 2013. In the same period of time, the area devoted to natural and cultivated pastures that changed from 1´570,000 to 4´486,000 ha. This clear expansion trend significantly reduced the availability of good new land and has opened the way to use páramos and marginal lands for agricultural purposes with the consequent environmental impact. The existence of altitudinal climatic zones and the heterogeneity of soils have developed a complex assembly of different situations that varies from areas that can be used intensively to areas that are not suitable for agricultural production. Crops like rice, a staple food for the Ecuadorian population, and corn for animal nutrition, are mail crops cultivated in the low southern coast plane, while the traditional perennial plantation crops like banana and oil palm are mainly cultivated on the central and northern coast plane; however, the area under cocoa cultivation has increased steadily all over the coast plane. Crop production in the Highlands is dominated by potato, open pollinated corn used mainly for human consumption, and horticultural crops along with extensive areas utilized for pastures for milking cows. Extensive areas of soils on the Amazonia have been incorporated to pasture to maintain cattle production, and more recently to corn, cocoa, and coffee production.

Keywords

Climatic zones • Population growth • Land pressure • Erosion • Cangahua

As a consequence of land use, soils from the Highlands, are those that have suffered a severe erosion process, mainly driven by rainfall, which have promoted the loss of the surface soil horizon exposing poor quality lower strata where it is difficult to conduct profitable agriculture. Erosion processes are also directly related to population density, responsible for the continuity of the process. Population growth and the consequent need to produce food promotes the development of intensive agriculture, which is usually practiced without implementing effective methods for soil protection to prevent the loss of topsoil, rich in nutrients and organic matter. The lack of effective implementation of soil conservation practices significantly affects the sustainability of this natural resource in Ecuador.

6.1 Land Use Trends

Given the existence of altitudinal climatic zones and the heterogeneity of soils (physical and chemical properties), soil agricultural potential varies from areas that can be used intensively to areas that are not suitable for agricultural production. Population growth, pressure on the land, and food and exportable products requirements has expanded the country agricultural frontier. The area under cultivation was

J. Espinosa (✉)
Facultad de Ciencias Agrícolas, Universidad Central del Ecuador, Quito, Ecuador
e-mail: aespinosa@uce.edu.ec

J. Moreno
Instituto Espacial Ecuatoriano, Quito, Ecuador
e-mail: julio.moreno@institutoespacial.gob.ec

J. Espinosa et al. (eds.), *The Soils of Ecuador*, World Soils Book Series,
https://doi.org/10.1007/978-3-319-25319-0_6

1′766,000 ha in 1954, while in 2013 the area increased to 2′ 470,000 ha. Similarly, it was also noteworthy the increase of almost threefold, in the same census periods, of the area devoted to natural and cultivated pastures that changed from 1′570,000 to 4′486,000 ha (INEC 2013; Pacheco 2009). In the 1990 decade, about two-thirds of the country was covered by natural vegetation (Fig. 1.15), but the Coastal Plane had lost at that time more area under this cover and only about 36% of the region remained protected by this type of vegetation, the rest was replaced mainly by plantations crops. On the other hand, in the Amazonia, only 2% of the land was used for annual crops, pastures, and permanent crops. In the Galapagos Islands, areas under agricultural exploitation reached only 2%. In the Highlands, annual crops covered about half of the area (Huttel et al. 1999). Recent census information indicates that these values have increased, for example, in the Amazonia, managed areas are now 8%, mostly with cultivated grasses. In the Coastal Plane, natural vegetation cover is 30%, and in the Highlands, the cultivated area increased from 30% in 2000 to 40% in 2013 (INEC 2013; Pacheco 2009). This shows the clear expansion trend of the agricultural frontier in Ecuador which has exhausted the possibility to have access to good land and opens the way to use páramos and marginal lands with the consequent environmental risk (FLACSO et al. 2008; IGM 2013).

6.1.1 Main Crops in the Coastal Plane

Main short cycle crops in the Coastal Plane are corn, and rice used as animal feed and as staple food, respectively. The main plantations crops are banana, oil palm, cocoa, coffee, and sugarcane which are cultivated in extensive areas of the region. A description of these main crops and other cultivated in smaller areas is as follows.

Rice (*Oryza sativa* L.)
Rice cultivation covered an area of 188,000 ha in 1991 (Huttel et al. 1999). In 2013, Ecuador harvested 396,770 ha of rice, making the country self-sufficient in this staple food (INEC 2013). This is a crop that needs clayey soils with good water retention.

Corn (*Zea mays* L.)
This crop, used for animal feed, is located in the low and warm plains of the Coastal Plane. In 1992, it occupied an area of 315,000 ha (Huttel et al. 1999) planted with varieties and hybrids. In 2013, the harvested area of corn in the region was 322,590 ha, planted with high yielding hybrids, at the moment the country is self-sufficient in corn (INEC 2013). Corn needs rich soils and plenty of water throughout the growing season, especially in the flowering season.

Banana (Musa sp div.)
This is the typical crop of the Coastal Plane which reached the highest production area in the 1950s and 1960s. In 1966, banana production covered an area of 163,000 ha (Huttel et al. 1999). In 2013, census data reported the existence of 160,920 ha in production (INEC 2013), most of the banana fruit production is for external markets. This crop requires homogeneous, deep, loamy, rich in potassium and nitrogen soils, and high water availability. Best production is achieved in areas with rainfall ranging from 120 to 150 mm/month, or in areas under irrigation. Banana tolerates varied sun exposure, but the demand for water increases with the number of hours of sunshine per year.

Oil Palm
Includes the *Elaeis guineensis* African species and the inter-specific hybrid O × G, result of the combination of native or American palm (*Elaeis oleifera*) with oil palm (*E. guineensis*). Oil palm cultivation in Ecuador started in 1950, and the area covered by the crop has been steadily growing, currently reaching an extension just over 280,000 ha (ANCUPA 2014). In general, oil palm is not a soil demanding crop, but grows best loam, clay loam, and silt loam soils with adequate drainage. This is a crop that demands high temperatures, ranging from 26 to 29 °C, but it also requires plenty of water (2000 and 2500 mm evenly distributed over the year), and sufficient light (1,000 or more hours/light/year).

Cacao (*Theobroma cacao* L.)
This is also an emblematic crop from the Coastal Plane which originated an evoked economic bonanza for this region at the beginning of the twentieth century. This bonanza, product of the cacao trade at that time was the end result of the influence of several factors that released the trade, such as the reduction of the Acapulco tax (1774), the Bourbon reform (1789), and the Independence of Guayaquil (1820). All of these factors freed the trade from the obligatory transit for Lima-Callao or Spain and encouraged the use of new land for perennial crop cultivation such as cocoa, which significantly increased cacao exports from 5000 tons in 1855 to 20,000 in 1900 and 47,000 in 1914. This way, cocoa cultivation was the main engine of the economy as far has 1940, when it was substituted by banana (Huttel et al. 1999). In 1999, cacao covered an area of 300,000 ha, the majority unwept plantations that lost importance due to the cacao price collapse during the 1920s. In 2013, cacao was produced in 301,551 ha at the Coastal Plane and 33,700 ha at the Amazonia (INEC 2013), a good portion on renovated plantations with increased production to take advantage of the global price increase, again making cacao cultivation a good contributor to the economy. Cacao is a crop that requires well-drained and fertile soils, warm climate,

temperatures above 10 °C and 1250–1500 mm of rainfall, but can withstand dry periods if these are not intense or prolonged.

Coffee (*Coffea* sp.)

Coffee has been cultivated in Ecuador since the eighteenth century. Statistical reports indicate that in 1999 coffee covered an area of 400,000 ha (Huttel et al. 1999) mostly with subsistence plantations of the Robusta variety planted in between the tropical forest with little technology. Official information indicates that in 2012 Ecuador harvested around 149,000 ha of coffee (PROECUADOR 2013), including well-managed areas of the Robusta and Arabica varieties that have substantially increased production for export purposes. The coffee is not very demanding in soil fertility and it can be cultivated in fertile soils as well as in poor soils even in acidic soils. The climatic requirements vary depending on the species, for instances the Arabic variety, native of eastern Africa, is grown at altitudes between 1300 and 2000 m, with rainfall ranging from 1500 to 1800 mm and temperatures from 20 to 25 °C, and a distinctive dry season. The Robusta variety, native to the humid forest of Africa, demands a warmer climate, >1500 mm of rainfall, temperatures from 22 and 26 °C and no more than 2 months of dry season.

Sugarcane (*Saccharum* sp.)

Introduced to Ecuador in the early sixteenth century, sugarcane is cultivated on the central plains of the Coastal Plane for about a century, condition which promoted the development of large industrial sugar mills. Reports indicate the existence of 90,361 ha of sugarcane in the Coastal Plane in 2013 (INEC 2013). This is a crop that is also planted in the Inter-Andean warm valleys of the Highlands. Sugarcane prefers friable soils, but tolerates different conditions if soils are deep, well drained and rich in organic matter. The sugarcane requires a hot and humid climate, and 100–120 mm of monthly rainfall during the vegetative growth phase, amount that must be supplied with irrigation in dry areas. The region intense insolation favors the accumulation of sugar in the stems and therefore the production.

Plantain (*Musa paradisiaca*) (genome AAB)

The plantain fruit requires cooking for consumption and is a staple food for population in the Coastal Plane and the Amazonia regions. However, certain areas produce surplus fruit that is exported to markets in the United States and Europe. In 2013, the reported area under plantain was 144,900 has, from which 86,700 ha are under monoculture and the rest are associated with other crops (INEC 2013).

Pineapple (*Ananas sativus* Schüles f.)

Traditionally, pineapple was a crop planted on the relatively dry areas from the Guayas river basin (especially around Milagro), as well as from El Oro province (Huttel et al. 1999). However, since the 1980s, pineapple cultivation extended to the humid areas of the country, particularly to Santo Domingo, where, in spite of the limited sunlight, the largest and most technologically advanced plantations are located. In 2012, Ecuador harvested 6700 of pineapple (FAO 2014).

Mango (*Mangifera indica* L.)

The areas under mango cultivation in Ecuador are located in the dry zones from the Guayas and Manabí provinces, where it occupied around 2700 ha in 1991; areas which were planted with local varieties and improved materials for export (e.g., Tommy Atkins) (Huttel et al. 1999). The recent growth of the area under technologically advanced mango plantations for export has been significant, reports indicate that 19,000 ha of this tropical fruit were harvested in 2012 (FAO 2014).

Cassava (*Manihot utilissima* C.)

The area cultivated with cassava in Ecuador has not changed substantially over the years, it is estimated that the actual area under cassava is around 20,000 ha (INEC 2013). Cassava is also one of the staple foods for the population, particularly in the Amazonia and Coastal Plane regions. Cassava is grown in areas with an annual rainfall pattern ranging from 500 to 3000 mm, mean temperatures of 23–24 °C and a dry seasons lasting no more than three months. This crop prefers slightly acidic soils and is demanding in potassium.

Rubber (*Hevea brasiliensis* M.)

Rubber has been grown in the Ecuadorian Coastal Plane since the first half of the twentieth century (Huttel et al. 1999; Patiño 1967). The area under rubber has increased gradually over the last years due to the renovated global interest on natural rubber for different purposes. In 2013, it was reported the existence of an area close to 12,000 ha of rubber to produce natural latex (FAO 2014). This crop requires deep, well-drained soils, and it grows best on rich and flat land. It requires warm and humid climate with plenty of sunlight, and a minimum annual rainfall of 1500 mm distributed evenly throughout the year. The production can be affected by the presence of dry periods.

Abaca (*Musa textiles* N.)

Abaca is mainly cultivated to obtain the fiber to manufacture waterproof paper. The first abaca plantations were established in the Ecuadorian Coastal Plane in the early 1950, during the intensive colonization of the humid areas of what is now the Santo Domingo de los Tsáchilas province, reaching an area of around 10,000 ha (Huttel et al. 1999). In 2012, the area under abaca cultivation was 25,040 ha (FAO

2014). This crop demands well-drained, rich soils and sufficient rainfall distributed throughout the year.

Soybean (*Glycine max* L. Merrill)
Soybean production in Ecuador started in 1971 to meet the country demand of edible oil. In 1992, an area of 120,000 ha under soybeans was reported, planted mainly in the Guayas and Los Ríos provinces (Huttel et al. 1999). Soybean area has gradually decreased over the years due to the high production costs which cannot compete with soybean production in other countries. In 2012, Ecuador harvested 30,000 ha of soybean (INEC 2013) mostly grown on soils of volcanic origin or recent alluvial deposits. This crop requires plenty of water (although it does not support water excess) and high temperatures.

Cotton (*Gossypium hirsutum* L.)
Cotton was grown in the Coastal Plane as early as the seventeenth century; in 1999 it was reported the existence of an area of about 30,000 ha planted with cotton (Huttel et al. 1999). Disease and insect problems made cotton production costly and, for this reason, the area harvested of cotton seed in 2012 was only 3250 ha (FAO 2014). This crop was also planted on the warm Inter-Andean valleys like El Chota where it remained until 1945. Cotton demands deep, permeable and rich soils, being tolerant to high salt concentrations.

6.1.2 Main Crops of the Highlands

Agriculture in the highlands is characterized mostly for production in small farms, with the exception of certain areas under livestock production. However, this production is important to satisfy food demand and to preserve food security for the population. A description of the main crops of the region is as follows.

Corn (*Zea mays* L.)
Corn is grown in the Highlands intermediate altitudinal strata using soft grain open pollinated varieties intended for human consumption. This is a staple food rooted in the tradition of the population. In 1992, the area cultivated with soft corn for dry grain was 187,310 ha and 21,580 ha for corn cob (Huttel et al. 1999). In 2013, the harvested area for dry grain was 79,519 ha, and 55,157 ha for corn cob (INEC 2013).

Potato (*Solanum tuberosum* L.)
This is one of the more extensive crops in the Highlands and constitutes a staple food not only for the rural population of the country. Potato production is in the hands of small farmers (<5 ha) as a traditional crop native from the Andes (Huttel et al. 1999; Pumisacho and Sherwood 2002). Potatoes are grown all over the Highlands, but especially in the Carchi, Chimborazo, Tungurahua, Cotopaxi and Pichincha provinces at altitudes from 2,800 to 3,200 m. The area under potato cultivations has not changed during the last few years, in 2013, the recorded potato area was 47,302 ha (INEC 2013). The potato grows well in deep, black, slightly acidic volcanic soils.

Barley (*Hordeum vulgare* L.)
Barley seed arrived to Ecuador in the hands of the Spaniard conquerors, but it was adopted by the rural population who included the grain into its diet. Since the mid-twentieth century, a high proportion of harvested grain was used for beer production. In the period 1985–1986, the barley planted area was 68,000 ha (Huttel et al. 1999), but this area has reduced significantly due to available imported grain at lower prices than the locally produced grain, for this reason, in 2013 only 26,300 ha of the grain were harvested (INEC 2013). Barley cultivation is concentrated in the provinces of Chimborazo, Cotopaxi, Pichincha, Bolívar, and Imbabura. This crop can adapt to conditions of high altitude (3750–3800 m), but it grow best at altitudes below 3600 m. Barley stands cold climate and frost while not eared. Barley is also grown in the dry zones of the intermediate Highland strata (2800–3000 m), with low rainfall and no irrigation water (500–700 mm).

Wheat (*Triticum sativum* L.)
This plant also arrived to Ecuador in the hands of the Spaniards, and its considered that it could have been planted for the first time in the San Francisco Plaza in Quito in 1541 (Patiño 1967). Even though the use of wheat flour for diverse culinary purposes, mainly bread preparation, is rooted among the Ecuadorian population, local wheat production has not been able to compete with external production and most of the consumed grain is imported. In 2012, only 9318 ha of wheat were harvested in Ecuador (FAO 2014). This is a crop that grows well at altitudes from 2200 to 3100 m, even though it can be also found at higher altitudes (3300 m). Wheat cultivation is concentrated in the provinces of Bolívar, Chimborazo, Imbabura, Pichincha, and Carchi, in soils with good moisture holding capacity that are not compacted in dry season, and receive between 600 to 700 mm of precipitation and 1050 light hours per year. Wheat grows in sites with temperatures of 5–6 °C, but can develop in areas up to 28 °C (Huttel et al. 1999).

6.1.3 Main Crops of the Amazonia

Due to the characteristic temperature, precipitation, and soil, virtually all tropical crops found in the Ecuadorian coast can be grown in the Amazonia, but cultivation areas are small. However, it is interesting to consider the increase in the cocoa (*Theobroma cacao* L.) planted area. In 2004, only 8,500 ha were cultivated with cocoa, while in 2013 the area increased to 32,000 ha (INEC 2013), indicating that this crop is an important economic commodity for the Ecuadorian Amazonia.

6.1.4 Forests and Pastures

The dense evergreen forest from the two great jungle massifs located in the Coastal Plane (Esmeraldas) and in the Amazonia (provinces of Napo and Sucumbíos) have been for a long time the source of the wood used by the industry, even knowing that wood exploitation causes severe environmental damage. The Esmeraldas forest is the most important for its particular characteristics, but the Amazon rainforest is much bigger and more complex. In addition, in the outer slopes of the Andes there exist forest bodies where species as cedar (*Cedrela*) or rosemary (*Podocuarpus*) are exploited, and in the dry zone there are compact forests where species such as guayacán (*Tabebuia* div. Sp.), orbeldaco (Bombax), are also harvested.

Cattle pasture and grasslands are present in all the country climatic conditions in large homogeneous areas. There are pastures in the Highlands, in the middle and lower parts of the wet external flanks of the Andes, and in the lower and warmer parts of the country. However, it is noteworthy the threefold increment of the area under pasture and grassland which passed from 1′570,000 ha in 1954 to 4′486,000 ha in 2013 (INEC 2013; Pacheco 2009). From the Amazonia 13 million ha, 2013 data estimated that 1’001,800 ha had human intervention, from which 797,200 ha were cultivated grasses, 131,300 ha permanent crops, 48,000 ha annual crops, and 25,400 ha fallow land (INEC 2013). This represents 8% of the Amazonia, an area that has increased in comparison to the 1990s decade when human intervened areas accounted for only 2% of the area (Huttel et al. 1999).

6.1.5 Secondary Crops

In addition to the main crops presented in the above discussion, a number of secondary crops are cultivated in Ecuador as a result of the diverse environmental conditions of the production systems. This diversity is due to the variety of soils and climates, as well as the long and asymmetrical evolution of agricultural land ownership and changing local population densities, conditions which result on virtually empty areas, usually occupied by forest, and areas heavily populated dedicated to agriculture (Huttel et al. 1999). Similarly, differences among large holdings (haciendas and plantations) and peasant smallholdings translate into major social and economic differences based on the land tenure structures, conditions that also promote the diversity of secondary crops.

Among the main secondary food crops are those native from the Andes like chocho (*Lupinus mutabilis* S.), melloco (*Ullucus* HBK), oca (*Oxalis tuberosus* Mol.), Quinoa (*Chenopodium quinoa* W.), and white carrot (*Arracacia zanthorrhiza* B.), all of them rich in minerals and amino acids. Crops like faba (*Vicia faba* L.) and green peas (*Pisum sativum* L.), both native from Middle East, are also important secondary crops. Part of the production of these crops is for household consumption, but an important remnant goes to local markets (Huttel et al. 1999) with the exception of quinoa that is mostly cultivated for export to diverse global markets (Peralta 2009).

On the other hand, 27 vegetables and legumes are grown in Ecuador, in small open plots or greenhouses mainly in the cold areas of the Highlands, along with 29 species of fruit trees in the Highlands, and 26 species in the warm areas of the Highlands and the Coastal Plane. Seventeen species of other secondary crops and 42 species of flowers (roses, summer flowers and flowers or tropical plants) are also cultivated in the country. It should be noted the significant increment of flower production in the Highlands, which has reached an area of 3500 ha in recent years, mostly grown in greenhouses with high technology (PROECUADOR 2012).

6.2 Soil Erosion

Of the three natural regions of continental Ecuador, soils from the Highlands, due to their topographic conditions, are those that have suffered a severe erosion process, mainly driven by rainfall, which have promoted the loss of the surface soil horizon exposing poor-quality lower strata where it is difficult to conduct profitable agriculture. The areas where erosion processes are evident are shown in Fig. 6.1. Erosion processes are also directly related to population density, responsible for the continuity of the process. Population growth and the consequent need to produce food promotes the development of intensive agriculture, which is usually practiced without implementing effective methods for soil protection to prevent the loss of topsoil, rich in nutrients and organic matter. The lack of effective implementation of soil conservation practices significantly affects the sustainability of this natural resource in Ecuador (de Noní et al. 1997; Espinosa 2014; Espinosa and Maldonado 1987).

6.2.1 Causes of Erosion

Rural population growth forced farmers to use marginal land for production, particularly land on steep slopes. For the same reason, farmers have divided the land developing a

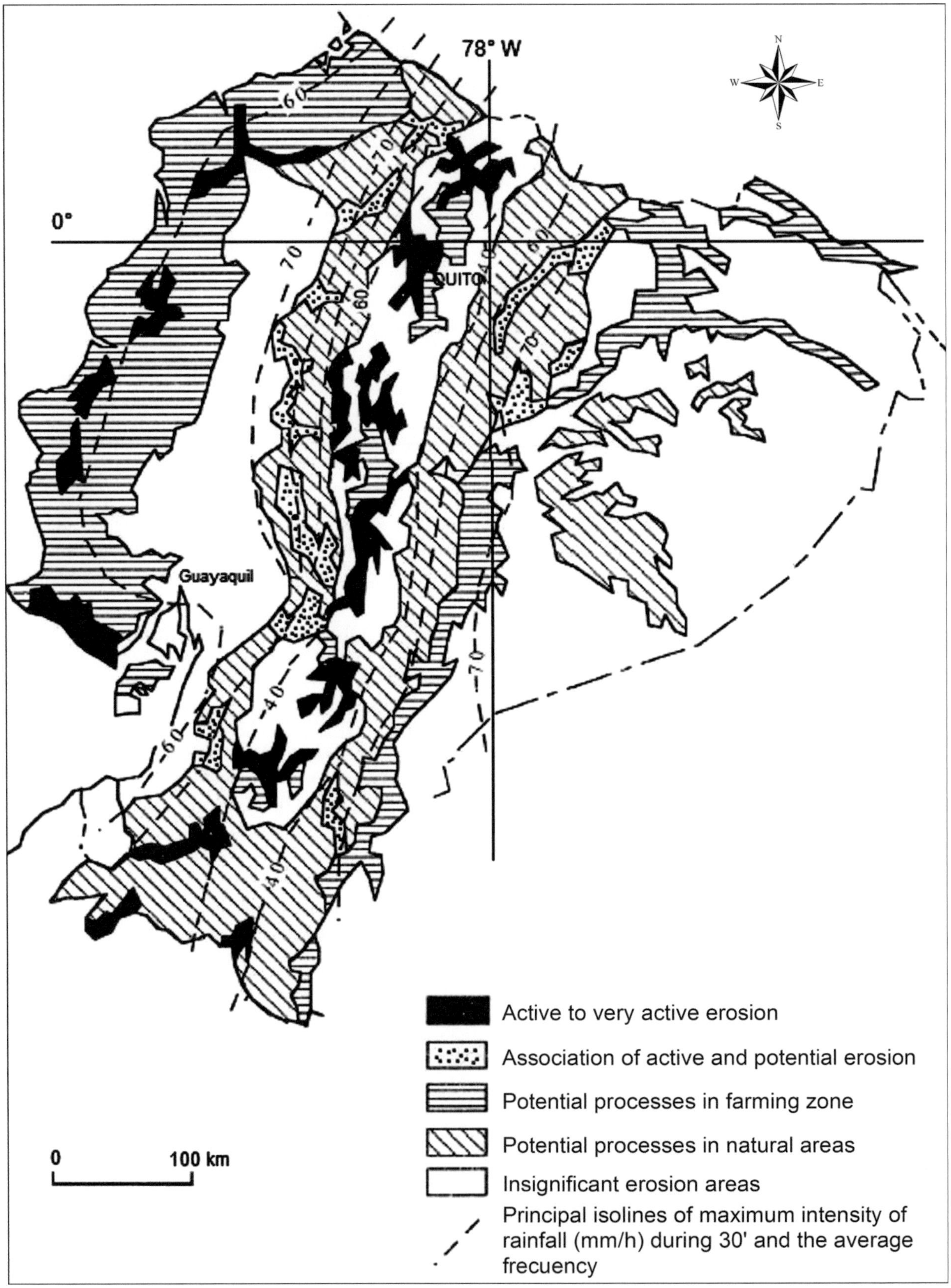

Fig. 6.1 Areas affected and activity degree of the erosion processes in Ecuador (de Noní et al. 1997)

Fig. 6.2 Distribution along the slope of small fields on steep land in the provinces of Chimborazo (left) and Tungurahua (right)

complicated network of small plots where farming is limited. This land division left long narrow fields leaning down the slope, and this land arrangement promotes erosion (Fig. 6.2) The limited availability of land for agriculture also promotes an intense land use and agricultural practices that promote soil loss. This is the case of plowing the soil in the direction of the gradient either with oxen or tractor (Fig. 6.3).

Maintaining soil surface without protection is a major cause of erosion in these zones. There are several reasons why farmers maintain soils bare; one of them is the use of crop residues as fuel or livestock feed. The agriculture conducted on these small plots is precarious due to the lack of resources and technology; crops grow poorly and unevenly and do not cover the soil or leave enough residues. Add to this, the downhill arrangement of the furrows, situation that only adds the problem (Dercon et al. 2007; Espinosa 2014).

It is necessary to make clear that the erosion does not belong only to small farmers; erosion is present with the same intensity on large poorly managed farms. Irrigation has contributed actively to the erosive process, in both small and large farms. Paradoxically, water needed for production ends up eliminating the fertile top soil layer, the main support for agricultural production. Very few irrigation projects include the indispensable soil conservation practices that would make efficient and sustainable the investment in irrigation (Southgate and Whitaker 1994; Vanacker et al. 2003).

6.2.2 Problems Caused by Erosion

The main consequence of erosion is the loss of productive soil, the most valuable resource for agriculture. Erosion eliminates the surface layers from the soil profile exposing new layers of low fertility and poor physical conditions; consequently, crops planted on these eroded areas have poor vigor and low productivity. As erosion progresses, the situation becomes more precarious until a completely

Fig. 6.3 Downhill land preparation on small farm fields in the provinces of Chimborazo (left) and Tungurahua (right)

Fig. 6.4 Slopes under severe erosion process in producers on small farms in the province of Chimborazo (left) and areas completely eroded with exposed cangahua in the province of Cotopaxi (right)

unproductive matrix emerges (Fig. 6.4). These conditions are common in many areas of the Ecuadorian Highlands bringing only poverty, leading finally to farmer abandonment of the land (Vanacker et al. 2003).

Soil erosion causes irreparable problems at sites where it occurs. Furthermore, soil particles leave farmer fields with water runoff along with plant nutrients and organic matter that farmers need to maintain in ground. The nutrients incorporated in the form of mineral and organic fertilizers leave the field where they are needed and accumulate elsewhere causing environmental problems. This way, soil particles are transported in the form of sediments to distant places reaching lower areas, in many cases other farmer fields, roads, and reservoirs. The accumulation of sediments on the roads is a serious problem for local municipalities; clearing sediment from roads is a troublesome and costly operation.

The sediment also reach important infrastructure like large reservoirs and dams in the lowlands. The large dams for energy generation lose capacity due to sediment accumulation coming mainly from farmer fields in many cases far away from them. The most serious problem occurred at the Paute dam, which nearly filled with sediments produced limited amounts of energy during the dry season (Donoso 2009; Harden 1993). The same could happen with new dams under construction.

Nutrient-laden sediments deposit in water bodies (rivers, lakes, and dams) and promote abundant growth of algae and other aquatic plants. These nutrients that could support vigorous crop growth on farmer field now promote the active growth of these aquatic plants in water bodies obstructing navigation and killing fish due to water eutrophication. In addition, the large volumes of accumulated plant material clog infrastructure for irrigation projects. Again, cleaning accumulated material is very costly and difficult.

6.2.3 Erosion Control

All agricultural properties in Ecuador, large or small, should have a definite soil conservation plan to prevent erosion. This way, production problems in the fields and the consequent sequel of indirect problems caused by erosion are avoided. In the past, in many places in the Highlands the problem of erosion was minimized because soils were deep and gullies ditches were covered with a pass of tractor or oxen leaving the sensation that nothing had happened. This has been a longstanding problem that has completely degraded soils in many places of the Sierra.

The methods to control erosion are well known; however, many producers see them as a nuisance because they require different land management from the traditional way which apparently is simple and effective to produce a crop. The objectives of these conservation practices are to protect soil surface and reduce the volume and velocity of runoff on the field. There are a few examples of soil conservation work on steep slopes in the Ecuadorian Highlands that successfully used mechanical and agronomic practices for erosion control (Fig. 6.5). Unfortunately, this type of projects is not common in extensive areas the Ecuadorian Highlands.

6.2.4 Cangahua

The cangahua is a hard subsurface stratum formed by volcanic tuffs that can be up to 100 m in depth located between 2000 and 3200 m a.s.l. at the internal Inter-Andean Flanks. These volcanic tuffs were scattered in what are now the provinces of Carchi, Imbabura, Pichincha, Cotopaxi, Tungurahua and Chimborazo in the Northern Highlands, covering an area of approximately 240,000 ha. These layers of tuffs were then covered by volcanic ash material that with

Fig. 6.5 Conservation management for erosion control on small farm fields in the province of Tungurahua

time developed fertile soils. The volcanic tuffs were hardened by the action of cementing agents such as carbonates and by the filling of the pores with clays, silica, and iron oxides. These conditions, combined with dry climatic periods, promoted the hardening of the material (Hall and Mothes 1997; Prat et al. 2015; Zebrowski 1997).

Pressure on land, steep slopes and poor soil management promoted the severe erosion of the surface soil derived from volcanic ash exposing the cangahua (Fig. 6.6). An estimation made in 1995 indicated that the area with exposed cangahua was around 80 000 ha located in areas of smallholding in the Northern Highlands characterized for having high population density (de Noni and Trujillo 1999a, b; Zebrowski 1997). Probably this surface is at the moment much higher, but no studies have determined the extent of the actual problem.

The farmers are forced to rehabilitate cangahua, at least partially, for agricultural use but yields of crops cultivated on this low fertility material are dramatically low (Fig. 6.6 right). Rehabilitating cangahua for agriculture is a social imperative task that must be satisfied with technological innovation. This would be a positive contribution to promote efficient and environmentally friendly agriculture among farmers, promoting in this way food security of the population.

Additionally, management of these poor rehabilitated cangahuas can be a good carbon (C) sink playing, therefore, an interesting role in local C sequestration. It has been demonstrated in Mexico that management of hardened materials (tepetates, similar to cangahuas) has the potential to sequester up to 90 Mg of C ha^{-1} (Baez Perez et al. 2002).

There are reports of cangahuas rehabilitation conducted manually or mechanically using tractors (Trujillo and Arias 1997); however, agricultural recovery of these materials has not been satisfactory. Habilitated cangahuas are very low in fertility, and contain low amounts of organic C and nutrients, particularly nitrogen and phosphorus, a situation that does promote good crop growth to provide good cover and the cycle of degradation starts again ending up in the loss of the rehabilitated material.

It is also important to consider that the incorporation to agriculture of eroded zones with cangahua outcrop would reduce the pressure on land avoiding the use of the paramo (land over 3200 m a.s.l.) for agriculture. The paramo is an important reservoir of water that must be conserved for the benefit of the population of extensive zones of the Northern Highlands (Podwokewski and Poulenard 2011).

A new effort to rehabilitate cangahua and transform it into soil began in 2012 in a joint effort between the MAGAP, FAO and several local governments from the area of Cayambe at the Norther Highlands. The work started with the breaking of the cangahua with machinery to create a plant physical support and establish a sustainable farming system in a relatively short period of time if proper conservation practices are placed on the ground. The process began with the incorporation of organic residues and mineral fertilizers, planting legumes and grasses to be incorporated as green manure to promote the development of a real soil matrix. Then traditional crops and new higher value crops will be planted. It is hoped that with this procedure, rehabilitated cangahua will produce adequate yields of wheat and beans in 3 years and maize in 5 years. In addition, it is expected that this method of cangahua recuperation will capture between 40 and 90 t ha^{-1} of C in the same period of time (Prat et al. 2015).

6.2.5 Need for a National Erosion Control Program

Ecuador needs to develop immediately a National Soil Conservation Plan to control erosion managing the effects of

Fig. 6.6 Cangahua under soil derived from volcanic ash in the province of Carchi (photo Jaime Hidrobo left) and zones with cangahua outcrops in the province of Cotopaxi (right)

rain and slope. As already indicated, this can be achieved using simple and inexpensive mechanical and agronomic practices of erosion control. While all this is essentially possible, perhaps the main obstacle to overcome is to convince the government, private entities, and the same farmer, of the urgent need to develop a National Program of Erosion Control.

Over the years, voices and opinions have been heard clearly emphasizing the need to take action to control the serious problem of soil loss from the agricultural areas of the country. Acosta Solis was one of the first scientists who documented the nature of erosion in Ecuador and defined erosion measures (Acosta Solis 1952). Economy and population growth during the following decades expanded the agricultural frontier and increased the pressure on land triggering an aggressive erosion process. A study conducted by the Natura Foundation in 1981 on the environment situation of Ecuador showed that the main environmental problem at that time was soil erosion (Espinosa 1981). That same year, the National Institute of Agricultural Research (INIAP) developed and implemented a soil conservation project which conducted activity in the field until 1986. This project trained more than 500 technicians and professionals throughout the country. These individuals were to be the human basis to implement a national plan of Soil Conservation. At the same time, the Ministry of Agriculture (MAG) created the National Soil Conservation Program (PRONACOS) as an extension of PRONAREG. INIAP resumed later its conservation program at small scale in some watershed at Highlands (Cordova and Novoa 1995). PRONACOS disappeared as a public entity 2 years after its creation due to ministry reorganization. Several private institutions such as CARITAS also develop field activities to control erosion for several years. Over time there have been warns of different type, by different means, about the need to develop effective actions to control soil erosion in Ecuador (Espinosa 2008; Jiménez 2011; Ochoa et al. 2013; Podwokewski and Poulenard 2011), but so far there is no a National Soil Conservation Program in place and the clear and undeniable fact is that severe soil erosion continues degrading the soil at the Highlands of Ecuador.

Under current conditions, a National Soil Conservation Program would have abundant basic information on the magnitude of the erosive process compiled and published by the Ministry of Agriculture, Livestock and Fisheries (MAGAP) and the Center for Integrated Natural Resource Remote Sensing (CLIRSEN), today Ecuadorian Space Institute (IEE), through projects such as SIG-Tierras and others (MIDENA et al. 2013). All these tools would make it easier and effective the planning and activity of a National Soil Conservation Program for the country.

6.2.6 Final Comments on Erosion

Erosion has degraded the soil, reduced crop production, and has had a profound impact on the economy of small and large farmers, particularly in the Highland of Ecuador. However, the problem also has indirect consequences for people in urban areas. It is also true that there is not a clear perception of how to attack the problem and what concrete actions are need in the field regardless the fact that soil erosion control methods are known and available. In cases, it has been demonstrated the feasibility of controlling soil erosion, but this information is concentrated in few groups related to agriculture but does not reach farmer fields.

Farmers generally perceive that something is wrong in their field management. This perception comes from the fact that crop yields are getting lower and that the soil is increasingly difficult to handle. However, farmer reject suggestions of implementing soil conservation practices because they believe it complicates the normal management of the plot and are apparently expensive. Developing awareness of erosion control in sectors not involved directly with agriculture could be even more difficult. Many believe that they are not part of the problem; however, they are indirect recipients of the consequences caused by erosion in rural zones.

A National Soil Conservation Program requires considerable investment and their results will not be immediately visible, for these reasons, it has not obtained the support from political sectors in the past. It would seem that the political conditions in the country have changed in recent years generating a favorable environment to implement a project of this nature, which could begin with the existing structure in the MAGAP, with the support of other agricultural organizations and producer associations.

Finally, it is imperative a radical change in the approach is used by the agricultural extension programs conducted by various government and private agencies. Agricultural extension should be based on soil conservation, in other words, any agricultural action should only start after any potential soil erosion problem at the site has been solved.

References

Acosta Solís M (1952) Por la conservación de las tierras andinas: la erosión en el Ecuador y métodos aconsejados para su control. Publicaciones Científicas MAS, Quito, Ecuador

ANCUPA (2014) Estadísticas de Palma Aceitera en Ecuador. https://www.google.com/search?q=ancupa&ie=utf-8&oe=utf-8. Accessed 10 diciembre 2014

Báez Pérez A, Etchevers Barra J, Hidalgo Moreno C, Prat C, Ordaz Chaparro V, Núñez Escobar R (2002) Carbono orgánico y P Olsen en tepetates cultivados de México. Agrociencia 36(6):643–653

Córdova J, Novoa V (1995) Problemática, experiencia y enfoque sobre la erosión, manejo y conservación de suelos de ladera en el Ecuador. In: IICA (ed) Manejo de Cuencas Hidrográficas. IICA, Lima, Perú

de Noni G, Trujillo G (1999a) La erosión actual y potencial en Ecuador: Localización, manifestación y causas ORSTOM. Quito, Ecuador

de Noni G, Trujillo G (1999b) Los principales procesos erosivos en el Ecuador. Ministerio de Agricultura y Ganadería, Quito, Ecuador

de Noni G, Trujillo G, Viennot M (1997) Análisis histórico, social y económico de la cangahua en Ecuador. In: Zebrowski C, Prat C, Etchevers J, Arias H, Miranda M (eds) I Simposio internacional Suelos volcánicos endurecidos, vol 10. Especial. Terra, México, pp 503–514

Dercon G, Govers G, Poesen P, Sánchez H, Rombaut R, Vandenbroeck G, Loaiza J, Deckers J (2007) Animal-powered tillage erosion assessment in the southern Andes region of Ecuador. Geomorphology 87:4–15

Donoso M (2009) La cuenca del río Paute: diagnóstico y propuesta de manejo integral. Políticas de desarrollo agropecuario, PROMAS, Cuenca, Ecuador

Espinosa J (1981) El recurso suelo en el Ecuador. In: Carrera L (ed) Diagnótico de la situación del medio ambiente en el Ecuador, vol 1. Fundación Natura-USAID Technical Information Center Quito, Ecuador

Espinosa J (2008) Distribución, uso y manejo de los suelos de la Región Andina. Paper presented at the XI Congreso Ecuatoriano de la Ciencia del Suelo, Quito, Ecuador

Espinosa J (2014) La erosión en Ecuador, un problema sin resolver. In: Cáceres E (ed) Siembra, vol 2. Facultad de Ciencia Agrícolas. Universidad Central del Ecuador, Quito, Ecuador, pp 56–70

Espinosa J, Maldonado F (1987) Logros y problemas de la conservación del suelo. Paper presented at the Conferencia Usos Sostenidos de Tierra en Laderas, Quito, Ecuador

FAO (2014) FAOSTAT. http://faostat3.fao.org//faostat-gateway/go/to/compare/Q/QC/S. Accessed 9 Dec 2014

FLACSO, MAE, PNUMA (2008) Geo Ecuador 2008: Informe sobre el estado del medio ambiente. FLACSO - MAE - PNUMA, Quito, Ecuador

Hall M, Mothes P (1997) El origen y edad de la cangahua superior, Valle de Tumbaco, Ecuador. In: Zebrowski C, Quantin P, Trujillo G (eds) III Symposium Internacional Suelos Volcánicos Endurecidos. ORSTOM-UE-PUCE-UCE, Quito, Ecuador, pp 19–28

Harden CP (1993) Land use, soil erosion, and reservoir sedimentation in an Andean drainage basin in Ecuador. Mount Res Dev 177–184

Huttel C, Zebronski C, Gondard P, Bourliaud J (1999) Paisajes agrarios de Ecuador., vol V. Geografía Básica del Ecuador. Instituto Panamericano de Geografía e Historia, Quito, Ecuador

IGM (2013) Atlas geográfico de la República del Ecuador. IGM - SENPLADES, Quito, Ecuador

INEC (2013) Encuesta de superficie y producción agropecuaria continua Ecuador 2013. Dirección de Estadísticas Agropecuarias y Ambientales, INEC, Quito, Ecuador

Jiménez S (2011) Erosión del suelo en Ecuador. CTT-USFQ - Fundación Carolina. http://observatoriopoliticaambiental.org/categoria-indicadores/89-indicadores-de-cambio-climatico/71-la-erosiondel-suelo-en-el-Ecuador. Accessed 12 de Abril 2015

MIDENA, SENPLADES, IEE, Gobierno Autónomo Decentralizado de Pichincha (2013) Memoria técnica geopedológica del cantón

Cayambe. Proyecto: Generación de geoinformación para la gestión del territorio a nivel nacional escala 1: 25 000. IEE, Quito, Ecuador
Ochoa P, Fries A, Montesinos P, Rodríguez J, Boll J (2013) Spatial estimation of soil erosion risk by land-cover change in the Andes of southern Ecuador. Land Degrad Dev 10:1002–1005
Pacheco R (2009) El Ecuador: Recursos naturales agrícolas y del medio ambiente. Imprenta Colón, Quito
Patiño V (1967) Plantas cultivadas y animales dométicos de América equinoccial. Tomo III: Fibras, medicinas, misceláneas. Imprenta Departamental, Cali, Colombia
Peralta E (2009) La quinoa en Ecuador: Estado del arte. INIAP. http://www.iniap.gob.ec/nsite/images/documentos/ESTADO%20DEL%20ARTE%20QUINUA%202.pdf. Accessed 10 de enero 2015
Podwokewski P, Poulenard J (2011) Suelos de los páramos de Ecuador. In: Mena P, Castillo A, Flores S et al (eds) páramo. Abya Yala, Quito, Ecuador
Prat C, Moreno J, Hibrobo J, Trujillo G, Ortega C, Etchevers J, Hidalgo C, Báez A, Gallardo J (2015) Turning rocks into soils from the Ecuadorian Andes to the Mexican Transvolcanic Sierra. In: Romeo R, Vita A, Manuellini S, Zanini E, Freppaz M, Stanchi S (eds) Understanding mountain soils: a contribution from mountain areas to the International Year of Soils. FAO, Rome, Italy
PROECUADOR (2012) Análisis sectorial de flores. PROECUADOR. http://www.proecuador.gob.ec/wp-content/uploads/downloads/2012/01/PROEC-AS2011-FLORES.pdf. Accessed 10 de diciembre 2014
PROECUADOR (2013) Monitoreo de las exportaciones. PROECUADOR. http://www.proecuador.gob.ec/monitoreo-de-exportaciones/. Accessed 10 diciembre 2014
Pumisacho M, Sherwood S (eds) (2002) El cultivo de papa en el Ecuador. INIAP-CIP, Quito, Ecuador
Southgate D, Whitaker M (1994) Economic Progress and the Environment. One Developing Country's Policy Crisis. Oxford University Press, New York
Trujillo G, Arias J (1997) Productibidad de la cangahua rehabilitada en Ecuador: Ensayos agronómicos. In: Zebrowski C, Quantin P, Trujillo G (eds) III Simposio Internacional de Suelos Volcánicos Endurecidos. ORSTOM-UE-PUCE-UCE, Quito, Ecuador, pp 260–271
Vanacker V, Govers G, Barros S, Poesen J, Deckers J (2003) The effect of short-term socio-economic and demographic change on landuse dynamics and its corresponding geomorphic response with relation to water erosion in a tropical mountainous catchment. Ecuador. Landscape Ecology 18(1):1–15
Zebrowski C (1997) Los suelos con cangahua en el Ecuador. In: Zebrowski C, Quantin P, Trujillo G (eds) III Simposio Internacional de Suelos Volcánicos Endurecidos. ORSTOM-UE-PUCE-UCE, Quito, Ecuador, pp 128–138

Index

J. Espinosa et al. (eds.), *The Soils of Ecuador*, World Soils Book Series,
https://doi.org/10.1007/978-3-319-25319-0

MIX
Papier aus verantwortungsvollen Quellen
Paper from responsible sources
FSC® C105338

If you have any concerns about our products,
you can contact us on
ProductSafety@springernature.com

In case Publisher is established outside the EU,
the EU authorized representative is:
Springer Nature Customer Service Center GmbH
Europaplatz 3, 69115 Heidelberg, Germany

Printed by Libri Plureos GmbH
in Hamburg, Germany